Enhanced Recovery Methods for Heavy Oil and Tar Sands

Enhanced Recovery Methods for Heavy Oil and Tar Sands

James G. Speight, Ph.D., D.Sc.
University of Trinidad and Tobago

Enhanced Recovery Methods for Heavy Oil and Tar Sands

Gulf Publishing Company
2 Greenway Plaza, Suite 1020
Houston, TX 77046

10 9 8 7 6 5 4 3 2 1

Library of Congress Cataloging-in-Publication Data

Speight, J. G.
Enhanced recovery methods for heavy oil and tar sands / James G. Speight.
p. cm.
Includes bibliographical references and index.
ISBN-13: 978-1-933762-25-8 : (alk. paper)
ISBN-10: 1-933762-25-X (alk. paper)
1. Enhanced oil recovery. 2. Petroleum. 3. Oil sands. I. Title.
TN871.S665 2009
622'.33827--dc22
2009008232

Printed in the United States of America
Printed on acid-free paper. ∞
Editing, design, composition, and indexing by TIPS Technical Publishing, Inc.

CONTENTS

LIST OF FIGURES

LIST OF TABLES

PREFACE

The declining reserves of light crude oil have resulted in an increasing need to develop options to upgrade the abundant supply of known heavy oil reserves. In addition, there is considerable focus and renewed efforts on adapting recovery techniques to the production of heavy oil.

Over the past decade, the demand for crude oil worldwide has substantially increased, straining the supply of conventional (light) oil. Recent price increases have emphasized the need for consideration of alternative or insufficiently utilized energy sources, especially heavy crude oil to supplement short- and long-term needs. Heavy oil has been used as refinery feedstock for considerable time, usually blending with more conventional feedstocks, but has commanded lower prices because of its lower quality relative to conventional oil.

Obviously, differences exist between heavy oil and conventional oil, according to the volatilities of the constituents. When the lower boiling constituents are lost through natural processes after evolution from organic source materials, the oil becomes heavy, with a high proportion of asphaltic molecules and with substitution in the carbon network of heteroatoms such as nitrogen, sulfur, and oxygen. Therefore, heavy oil, regardless of its source, always contains the heavy fractions, the asphaltic materials, which consist of resins and asphaltenes (Figure 2–1). Removal or reduction of the asphaltene fraction, through deasphalting or leaving these constituents in the reservoir during recovery, improves the refinability of heavy oil.

The objective of this book is to present to the reader the current methods of recovery for heavy oil and tar sand bitumen technology by nonthermal and thermal methods. The book is designed to be suitable for undergraduate students, graduate students, and professionals who are working with heavy oil and tar sand bitumen. Each chapter will include a list of references that will guide the reader to more detailed information.

—Dr. James G. Speight
The University of Trinidad and Tobago
Point Lisas Campus, Couva
Trinidad
July, 2008

CHAPTER 1

DEFINITIONS

Petroleum (crude oil; conventional petroleum) is found in the microscopic pores of sedimentary rocks such as sandstone and limestone. Not all of the pores in a rock contain petroleum, and some pores will be filled with water or brine that is saturated with minerals. However, not all of the oil fields that are discovered are exploited; the oil may be far too deep or of insufficient volume or the oilfield may be so remote that transport costs would be high.

Heavy oil is a viscous type of petroleum that contains a higher level of sulfur than conventional petroleum that occurs in similar locations (IEA, 2005; Ancheyta and Speight, 2007 and references cited therein). The nature of heavy oil is a problem for recovery operations and for refining—the viscosity of the oil may be too high, thereby rendering recovery expensive and/or difficult, and the sulfur content may be high, thereby increasing the expense of refining the oil.

In any text related to the properties and behavior (e.g., recovery or refining) of a natural resource (e.g., heavy oil), it is necessary to understand the resource first through the name or terminology or definition. *Terminology* is the means by which various subjects are named so that reference can be made in conversations and in writings and so that the meaning is passed on. *Definitions* are the means by which scientists and engineers communicate the nature of a material to each other and to the world, through either the spoken or the written word. Thus the definition of a material can be extremely important and have a profound influence on how the technical community and the public perceive that material.

Because of the need for a thorough understanding of petroleum and the associated technologies, it is essential that the definitions and the terminology of petroleum science and technology be given prime consideration. This will aid in a better understanding of petroleum, its constituents, and its various fractions. Of the many forms of terminology that have been used, not all have survived, but the more commonly used are illustrated here. Particularly troublesome, and more confusing, are those terms that are applied to the more viscous materials, for example the use of the terms *bitumen* and *asphalt*. This part of the text attempts to alleviate much of the confusion that exists, but it must be remembered that the terminology of petroleum is still open to personal choice and historical usage.

The name *heavy oil* can often be misleading as it has also been used in reference to (1) fuel oil that contains residuum left over from distillation, i.e. residual fuel oil, (2) coal tar creosote, or (3) viscous crude oil. For the purposes of this text the term is used to mean *viscous crude oil*.

Heavy oil typically has relatively low proportions of volatile compounds with low molecular weights and quite high proportions of lower-volatility compounds with high molecular weights. The high molecular weight fraction of heavy oils is composed of a complex assortment of different molecular and chemical types, a mixed bag of compounds (not necessarily just paraffins or asphaltenes) with high melting points and high pour points that greatly contribute to the poor fluid properties of heavy oil. This contributes to its low mobility compared to conventional crude oil.

The mobility of reservoir fluids influences recovery rates, but the enhanced oil recovery and artificial lift methods needed to produce the fluids change the already complex fluid characteristics of heavy oil. In order to correctly specify the necessary downhole equipment, it is important to understand those fluid properties (Chapter 3 and 4) and how they might change throughout the system. For example, in Alaska's West Sak and Schrader Bluff formations, heavy oil viscosity ranges from about 30 to 3,000 centipoise (Taylor, 2006).

Heavy oil typically has low levels of paraffins (straight-chain alkanes), if any at all, with moderate to high levels of asphaltene constituents. The asphaltene constituents are not necessarily the primary cause for the high specific gravity (low API gravity) of the oil, nor are they always the prime cause for production problems. It is essential to consider the content of the resin constituents and the aromatic constituents, both of which are capable of hindering the asphaltenes constituents from separation during recovery. It is only when the asphaltene constituents separate from the oil as a separate phase that they deposit in the formation or in the production train.

1.1 History

Petroleum, in various forms, is not a recent discovery (Abraham, 1945; Forbes, 1958a, 1958, 1959, 1964; Speight, 1978, 2007; Totten, 2007). More than four thousand years ago, bitumen from natural seepages was employed in the construction of the walls and towers of Babylon. Ancient writings tablets indicate the medicinal and lighting uses of petroleum in various societies.

In terms of recovery, the earliest known wells were drilled in China in 347 BC to depths of 800 feet (240 meters) and were drilled using bits attached to bamboo poles. The oil was burned to evaporate brine and produce salt. By the 10th century, extensive bamboo pipelines connected oil wells with salt springs.

The use of petroleum in the Middle East was established by the 8th century, when the streets of the newly constructed Baghdad were paved with the nonvolatile residue derived from accessible petroleum and seepages (particularly Hit) in the region. In the 9th century, petroleum was distilled at Baku, Azerbaijan, to produce naphtha, which formed the basis of the incendiary *Greek fire* (Cobb and Goldwhite, 1995). These Baku experiences were reported by the geographer Masudi in the 10th century and by Marco Polo in the 13th century, who described the output of those wells as hundreds of shiploads.

The earliest mention of American petroleum occurs in Sir Walter Raleigh's documentation of the Trinidad Asphalt Lake (also called the Trinidad Pitch Lake) in 1595. In 1632, the journal of Franciscan Joseph de la Roche d'Allion, which described his visit to the oil springs of New York, was published in Sagard's *Histoire du Canada*. A Russian traveler, Peter Kalm, in his work on America published in 1748 showed on a map the oil springs of Pennsylvania.

In 1854, Benjamin Silliman, a science professor at Yale University in New Haven, Connecticut, followed some of the work by Arabic alchemists and fractionated petroleum by distillation. Discoveries such as this rapidly spread around the world, and the first Russian refinery was built in the then-mature oil fields at Baku in 1861; at the time about 90% of the world's oil was produced at Baku.

The first commercial oil well drilled in North America was in Oil Springs, Ontario, Canada, in 1858 by James Miller Williams. The U.S. petroleum industry began with Edwin Drake's drilling of a 69-foot (21-meter) oil well in 1859 at Oil Creek, near Titusville, Pennsylvania, for the Seneca Oil Company. The well originally yielded 25 barrels per day, and by the end of the first year, output was at the rate of 15 barrels per day. The industry grew through the 1800s, driven by the demand for kerosene and for oil lamps. Petroleum refining became even more popular, perhaps essential, in the early part of the 20th century with the introduction of the internal combustion engine, which provided a demand that has largely sustained the industry during the past 100 years. Early finds like those in Pennsylvania and Ontario were quickly outpaced by demand, leading to oil booms in Texas, Oklahoma, and California.

By 1910, significant oil fields had been discovered and were being developed at an industrial level in Canada, the Dutch East Indies (1885, in Sumatra), Iran, (1908, in Masjed Soleiman), Venezuela, and Mexico. Until the mid-1950s, coal was still the world's foremost fuel, but oil quickly took over. The 1973 and 1979 energy crises brought to light the concern that oil is a limited resource that will diminish, at least as an economically viable energy source. At the time, the most common and popular predictions were spectacularly dire.

Petroleum's uses as a portable, dense energy source powering the vast majority of vehicles and as the base of many industrial chemicals make it one of the world's most important commodities. Access to it was a major factor in several military conflicts, including World War II and the more recent wars in the Persian Gulf. Approximately 80% of the world's readily accessible reserves of conventional petroleum are located in the Middle East, with the majority in Saudi Arabia. However, when the reserves of heavy oil and tar sand bitumen are taken into account, the balance shifts. Venezuela and Canada have substantial reserves of heavy oil and tar sand bitumen that are sufficient to shift the balance of petroleum reserves from the Middle East to the Americas. It is to this subject that this book is devoted.

Although bitumen will receive some mention, as a point of reference and comparison, the focus of this book is on heavy oil and the means by which it can be recovered. It is necessary through definitions to understand the nature of heavy oil vis-à-vis petroleum (conventional crude oil) and tar sand bitumen.

Currently, the oil industry is in need of a stimulus (in spite of oil at a high of $147 per barrel during the summer of 2008, at the time of writing) since some of the most prolific basins (e.g., Mexico's Cantarell oil field) have begun to experience reduced production rates and are reaching or have reached maturity. At the same time, the world's demand for oil continues to grow every year, fueled in part by the rapidly growing economies of China and India. The declining availability of conventional oil combined with rising demand has driven up oil prices and put more pressure on the search for alternate energy sources.

The stimulus needed is in the form of the reservoirs of heavy oil that are found in the Western hemisphere. These resources are more difficult and costly to extract, so they have barely been touched in the past. However, with these resources, the world could soon have access to oil sources almost equivalent to those of the Middle East.

With the price of oil reaching new highs, investments in these more challenging reservoirs are rapidly accelerating. The worldwide importance of heavy oils will continue to emerge as the price of oil remains high and the demand for it remains strong. Although prices are expected to fluctuate, it is worth moving ahead with heavy oil resources on the basis of obtaining a measure (as yet undefined and country dependent) of oil independence.

1.2 Petroleum

Definitions of *petroleum* have been varied, unsystematic, and often archaic. In fact, there has been a tendency to define petroleum and heavy oil on the basis of a single property. While this may be suitable for a general understanding, it is by no means accurate and does not reflect the true nature or characterization of petroleum or heavy oil. Unfortunately, identification or differentiation by use of a single property is a product of many years of growth. Its long-established use, however general or inadequate it may be, is altered with difficulty, and a new term, however precise, may be adopted only slowly.

Petroleum is a naturally occurring mixture of hydrocarbons, generally in a liquid state, which may also include compounds of sulfur nitrogen oxygen metals and other elements (ASTM, 4175).

Thus, petroleum and the equivalent term *crude oil*, cover a wide assortment of materials consisting of mixtures of hydrocarbons and other compounds containing variable amounts of sulfur, nitrogen, and oxygen, which may vary widely in specific gravity, API gravity, and the amount of residuum (Table 1–1). Metal-containing constituents, notably those compounds that contain vanadium and nickel, usually occur in the more viscous crude oils in amounts up to several thousand parts per million and can have serious consequences during processing of these feedstocks (Speight, 1984, 2007). Because petroleum is a mixture of widely varying constituents and proportions, its physical properties also vary widely (e.g., the color varies from near colorless to black).

In the crude state, petroleum has minimal value, but when refined it provides high-value liquid fuels, solvents, lubricants, and many other products. Crude petroleum can be separated into a variety of different generic fractions by distillation. The terminology of these fractions has been bound by utility and often bears little relationship to composition.

The fuels derived from petroleum contribute approximately one third to one half of the total world energy supply and are used not only for transportation fuels (i.e., gasoline, diesel fuel, and aviation fuel, among others) but also to heat buildings. Petroleum products have a wide variety of uses that vary from gaseous and liquid fuels to near-solid machinery lubricants. In addition, the residue of many refinery processes, asphalt (a once-maligned by-product), is now a premium value product for highway surfaces, roofing materials, and miscellaneous waterproofing uses.

The molecular boundaries of petroleum cover a wide range of boiling points and carbon numbers of hydrocarbon compounds and other compounds containing nitrogen, oxygen, and sulfur, as well as metal-containing (porphyrin) constituents. However, the actual boundaries of such a *petroleum map* can only be arbitrarily defined in terms of boiling point and carbon number. In fact, petroleum is so diverse that materials from different sources exhibit different boundary limits, and for this reason alone, it is not surprising that petroleum has been difficult to map in a precise manner.

The proportions in which the different constituents occur vary with the origin and the relative amounts of the source materials that form the initial *protopetroleum* as well as the maturation conditions. Thus, some crude oils have higher proportions of the lower-boiling components

Table 1–1 Typical Variations in the Properties of Petroleum

Petroleum	Specific Gravity	API Gravity	Residuum >1000 F % v/v
US Domestic			
California	0.858	33.4	23.0
Oklahoma	0.816	41.9	20.0
Pennsylvania	0.800	45.4	2.0
Texas	0.827	39.6	15.0
Texas	0.864	32.3	27.9
Foreign			
Bahrain	0.861	32.8	26.4
Iran	0.836	37.8	20.8
Iraq	0.844	36.2	23.8
Kuwait	0.860	33.0	31.9
Saudi Arabia	0.840	37.0	27.5
Venezuela	0.950	17.4	33.6

and others (such as heavy oil and bitumen) have higher proportions of higher-boiling components (asphaltic components and residuum).

Petroleum occurs underground, at various pressures depending on the depth. Because of the pressure, it contains considerable natural gas in solution. Underground petroleum is much more fluid than it is on the surface and is generally mobile under reservoir conditions because the elevated temperatures (the *geothermal gradient*) in subterranean formations decrease the viscosity. Although the geothermal gradient varies from place to place, it is generally on the order of 25 to 30°C/km (15°F/1000 ft or 120°C/1000 ft, i.e. 0.015°C per foot of depth or 0.012°C per foot of depth).

Petroleum is derived from aquatic plants and animals that lived and died hundreds of millions of years ago. Their remains mixed with mud and sand in layered deposits that, over the millennia, were geologically transformed into sedimentary rock. Gradually, the organic matter decomposed and eventually formed petroleum (or a related precursor), which migrated from the original source beds to more porous and permeable rocks, such as *sandstone* and *siltstone*, where it finally became entrapped. Such entrapped accumulations of petroleum are called *reservoirs*. A series of reservoirs within a common rock structure or a series of reservoirs in separate but neighboring formations is commonly referred to as an *oil field*. A group of fields is often found in a single geologic environment known as a *sedimentary basin* or *province*.

The major components of petroleum are *hydrocarbons*, compounds of hydrogen and carbon that display great variation in their molecular structure. The simplest hydrocarbons are a large group of chain-shaped molecules known as *paraffins*. This broad series extends from methane, which forms natural gas, to liquids that are refined into gasoline, to crystalline waxes. A series of ring-shaped hydrocarbons, known as *naphthenes*, ranges from volatile liquids such as *naphtha* to high-molecular-weight substances isolated as the *asphaltene* fraction. Another group of ring-shaped hydrocarbons is known as the *aromatics*; the chief compound in this series is benzene, a popular raw material for making petrochemicals.

Nonhydrocarbon constituents of petroleum include organic derivatives of nitrogen, oxygen, sulfur, and the metals nickel and vanadium. Most of these impurities are removed during refining.

Geologic techniques can determine only the existence of rock formations that are favorable for oil deposits, not whether oil is actually there. Drilling is the only sure way to ascertain the presence of oil. With modern rotary equipment, wells can be drilled to depths of more than 30,000 feet (9,000 meters). Once oil is found, it may be recovered (brought to the surface) by the pressure created by natural gas or water within the reservoir. Crude oil can also be brought to the surface by injecting water or steam into the reservoir to raise the pressure artificially or by injecting such substances as carbon dioxide, polymers, and solvents to reduce crude oil viscosity. Thermal recovery methods are frequently used to enhance the production of heavy crude oils, whose extraction is impeded by viscous resistance to flow at reservoir temperatures.

Petroleum is typically recovered from the reservoir by the application of primary and secondary recovery techniques. Although covered elsewhere (Chapter 5), mention of primary, secondary, and tertiary recovery is warranted here in terms of the general description of these techniques.

Primary recovery refers to the process in which the petroleum in the reservoir trap is forced to the surface by the natural pressure contained in the trap. This pressure may result from several forces: (1) when the reservoir is penetrated, the pressure release allows the water layer to expand, push the oil upwards, and replace it in the rock pores (the most effective technique, known as a water drive system) (Figure 1–1); (2) if the drill penetrates into a layer of oil that has a gas cap above it, the release of pressure allows the gas layer to expand rapidly, causing a downward pressure on the oil, forcing it to move up through the well (gas cap drive system) (Figure 1–2), and (3) gas dissolved in the oil may be released as bubbles when the trap is pierced; as the oil moves up, the gas in the oil expands and the growing bubbles push the oil to the surface (solution gas drive system) (Figure 1–3). In most reservoir traps, these pressures are sufficient to initially force the petroleum to the surface.

At some point in time, the pressure will fall. Petroleum production decreases because: (1) there is less force driving the oil towards the well, (2) the gas that moves into the emptied pore spaces reduces the permeability of the rock, making it more difficult for oil to flow through, and/or (3) the fall in pressure and the loss of dissolved gas increases the surface tension and viscosity of the oil. Thus, primary recovery techniques usually account for less than 30% of the total volume of petroleum recovered.

Secondary recovery involves trying to maintain reservoir pressure. One technique is to inject natural gas into the reservoir above the oil, forcing the oil downwards, and then injecting water below the oil, forcing it upwards. Sometimes the gas that is used is that which has just been released during primary recovery. The disadvantage of using the released gas is that this gas is a marketable product in its own right. However, this is a good method to use if transporting the gas would be costly. In any case, the re-injected gas can always be collected again if necessary. Alternative secondary techniques involve injecting carbon dioxide or nitrogen into the oil. This makes the oil more fluid, and the gas pushes the oil upwards.

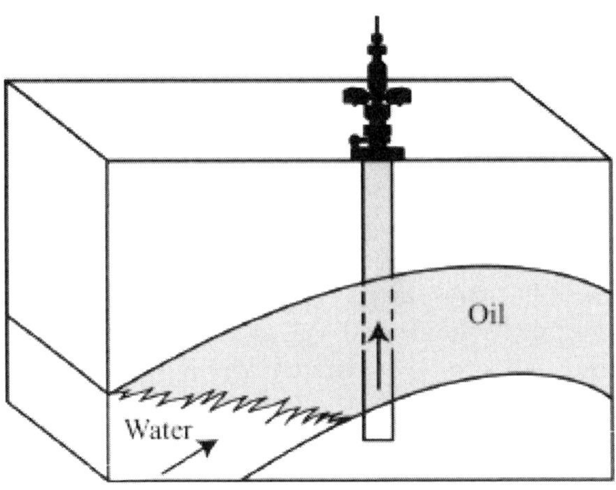

Figure 1–1 *Water drive.*

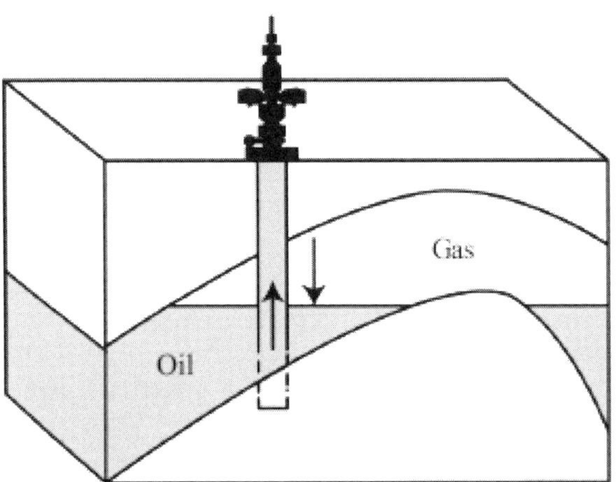

Figure 1–2 *Gas cap drive.*

Tertiary recovery is the most expensive approach and involves injecting steam, detergents, solvents, bacteria or bacterial nutrient solutions into the remaining oil. When high-pressure steam is injected, it heats the oil, decreasing its density and viscosity and increasing its flow rate (Figure 1–4). Sometimes, some of the oil in the reservoir rock is deliberately set on fire. This is done to increase the flow rate of the oil ahead of the combustion front. Detergents which

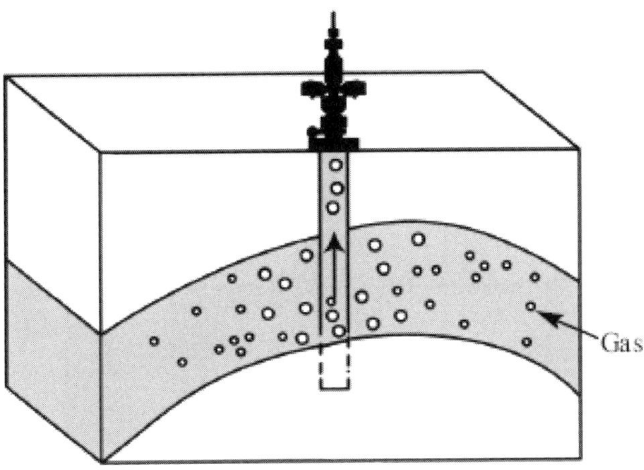

Figure 1–3 *Solution gas drive.*

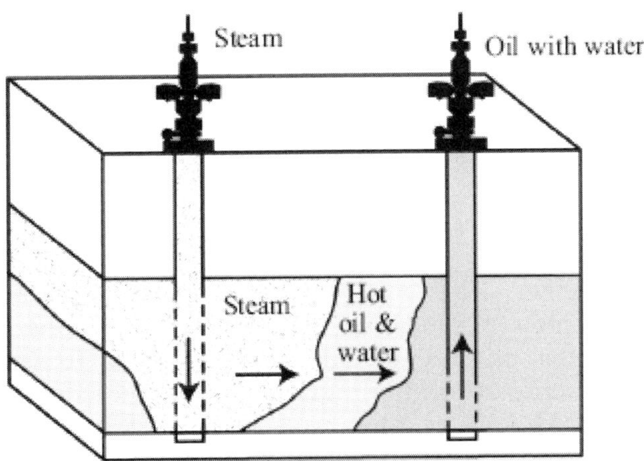

Figure 1–4 *Steamflooding.*

can be injected reduce the viscosity of the oil and act as surfactants, reducing the ability of the oil to stick to the rock surface and thus making it easier for it to be flushed up to the surface (Figure 1–5).

Another tertiary recovery technique involves injecting bacteria into the oil field. Some bacteria produce polysaccharides which reduce the

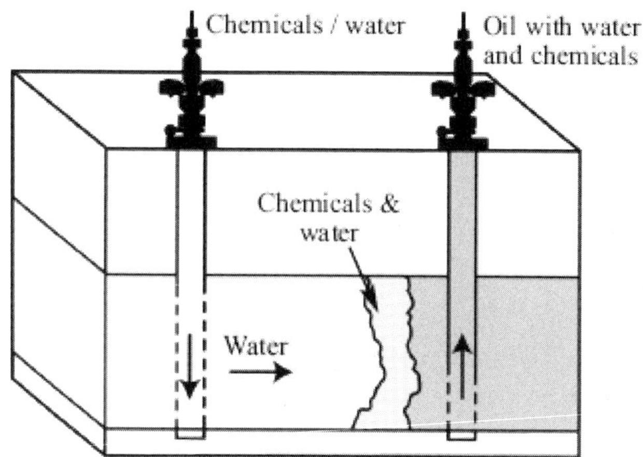

Figure 1–5 *Recovery using chemicals or detergents.*

permeability of the water-filled pores of the reservoir rock, and this effectively forces injected water into the oil-filled pores, pushing the oil out. Other bacteria produce carbon dioxide which helps to increase pressure within the rock pores, forcing out the oil. Other bacteria produce surfactants and/or chemicals that reduce the viscosity of the oil.

After recovery, petroleum is transported to refineries by pipelines, which can often carry more than 500,000 barrels per day, or by ocean-going tankers. The basic refinery process is *distillation*, which separates the crude oil into fractions of differing volatility. After the distillation, other physical methods are employed to separate the mixtures, including absorption, adsorption, solvent extraction, and crystallization. After physical separation into such constituents as light and heavy naphtha, kerosene, and light and heavy gas oils, selected petroleum fractions may be subjected to conversion processes, such as thermal cracking (i.e., coking) and catalytic cracking. In the most general terms, cracking breaks the large molecules of heavier gas oils into the smaller molecules that form the lighter, more valuable naphtha fractions.

Reforming changes the structure of straight-chain paraffin molecules into branched-chain *iso*-paraffins and ring-shaped aromatics. The process is widely used to raise the octane number of gasoline obtained by distillation of paraffinic crude oils.

1.3 Heavy Oil

There are large resources of *heavy oil* in Canada, Venezuela, Russia, the United States, and many other countries. The resources in North America alone provide a small percentage of current oil production (approximately 2%); existing commercial technologies could allow for significantly increased production. Under current economic conditions, heavy oil can be profitably produced, but at a smaller profit margin than for conventional oil, due to higher production costs and upgrading costs in conjunction with the lower market price for heavier crude oils.

Heavy oil is a type of petroleum that is different from conventional petroleum insofar as it is much more difficult to recover from the subsurface reservoir. It has a much higher viscosity (and lower API gravity) than conventional petroleum, and recovery of this petroleum type usually requires thermal stimulation of the reservoir. When petroleum occurs in a reservoir that allows the crude material to be recovered by pumping operations as a free-flowing dark- to light-colored liquid, it is often referred to as *conventional petroleum*.

Put simply, heavy oil is a type of crude oil that is very viscous and does not flow easily. The common characteristic properties (relative to conventional crude oil) are high specific gravity, low hydrogen-to-carbon ratios, high carbon residues, and high contents of asphaltenes, heavy metal, sulfur, and nitrogen. Specialized recovery and refining processes are required to produce more useful fractions such as naphtha, kerosene, and gas oil.

Heavy oil is an oil resource that is characterized by high viscosities (i.e. resistance to flow) and high densities compared to conventional oil. Most heavy oil reservoirs originated as conventional oil that formed in deep formations but migrated to the surface region, where they were degraded by bacteria and by weathering and where the lightest hydrocarbons escaped. Heavy oil is deficient in hydrogen and has high carbon, sulfur, and heavy-metal content. Hence, heavy oil requires additional processing (upgrading) to become a suitable refinery feedstock for a normal refinery.

Heavy oil accounts for more than double the resources of conventional oil in the world, and heavy oil offers the potential to satisfy current and future oil demand. Not surprisingly, heavy oil has become an important theme in the petroleum industry, with an increasing number of operators getting involved or expanding their plans in this market around the world.

However, heavy oil is more difficult to recover from the subsurface reservoir than conventional or light oil. A very general definition of heavy oil is based on the API gravity or viscosity. This definition is quite arbitrary, although there have been attempts to rationalize the definition based upon API gravity, viscosity, and density.

For example, heavy oils have been considered to be those crude oils that had gravity less than 20° API, with heavy oils generally falling into the API gravity range 10 to 15°. For example, Cold Lake heavy crude oil has an API gravity equal to 12°. Extra heavy oils, such as tar sand bitumen, usually have an API gravity in the range 5 to 10° (Athabasca bitumen = 8° API). Residua varies depending upon the temperature at which distillation is terminated, but usually vacuum residua are in the range 2 to 8° API (Speight, 2000, and references cited therein; Speight and Ozum, 2002, and references cited therein). In addition, heavy oils usually, but not always, have sulfur content higher than 2% by weight (Speight, 2000).

The term heavy oil has also been arbitrarily (incorrectly) used to describe both the heavy oils that require thermal stimulation of recovery from the reservoir and the bitumen in bituminous sand (tar sand) formations from which the heavy bituminous material is recovered by a mining operation. *Extra heavy oil* is a non-descript term (related to viscosity) of little scientific meaning that is usually applied to tar sand bitumen, which is generally incapable of free flow under reservoir conditions.

The methods outlined in this book for heavy oil recovery focus on heavy oils with API gravity of less than 20°; examples of such heavy oils are presented (Table 1–2). However, it must be recognized that some of these heavy oils are pumpable and are already being recovered by this method. Recovery depends not only on the characteristics of the oil, but also on the characteristics of the reservoir, including the temperature of the reservoir and the pour point of the oil (see also Chapter 4). These heavy oils fall into a range of high viscosity (Figure 1–6). The viscosity is subject to temperature effects (Figure 1–7), which is the reason for the application of thermal methods to heavy oil recovery.

Table 1–2 API Gravity and Sulfur Content of Selected Heavy Oils[a]

Heavy Oils	API	Sulfur wt.%
Bachaquero	13.0	2.6
Boscan	10.1	5.5
Cold Lake	13.2	4.1
Huntington Beach	19.4	2.0
Kern River	13.3	1.1
Lagunillas	17.0	2.2
Lloydminster	16.0	2.6
Lost Hills	18.4	1.0
Merey	18.0	2.3
Midway Sunset	12.6	1.6
Monterey	12.2	2.3
Morichal	11.7	2.7
Mount Poso	16.0	0.7
Pilon	13.8	1.9
San Ardo	12.2	2.3
Tremblador	19.0	0.8
Tia Juana	12.1	2.7
Wilmington	17.1	1.7
Zuata Sweet	15.7	2.7

a. For reference, Athabasca tar sand bitumen has API = 8° and sulfur content = 4.8% by weight.

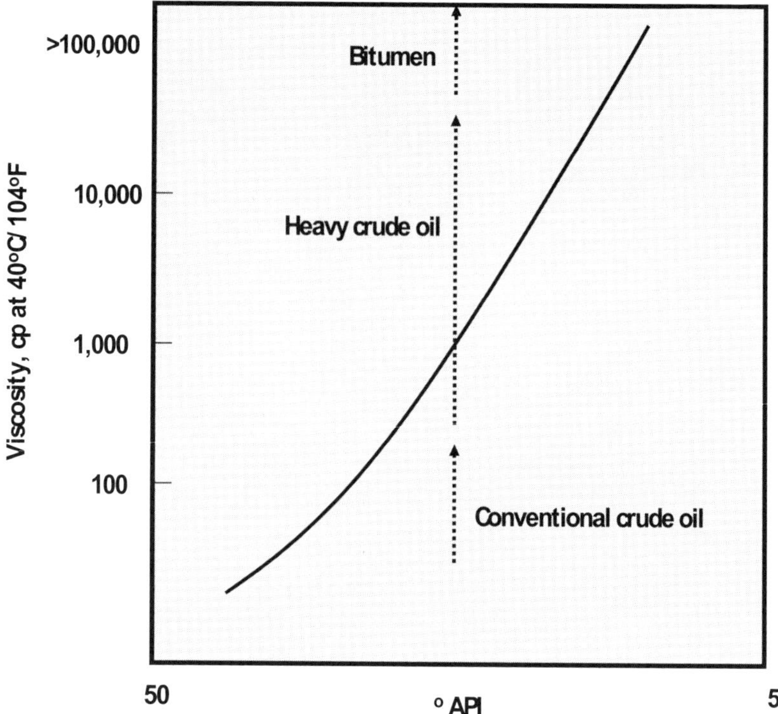

Figure 1–6 *General relationship of viscosity to API gravity.*

1.4 Tar Sand Bitumen

The term *bitumen* (also, on occasion, referred to as *native asphalt* and *extra heavy oil*) includes a wide variety of reddish brown to black materials of semisolid, viscous-to-brittle character that can exist in nature with no mineral impurity or with mineral matter contents that exceed 50% by weight. Bitumen is frequently found filling pores and crevices of sandstone, limestone, or argillaceous sediments, in which case the organic and associated mineral matrix is known as *rock asphalt* (Abraham, 1945).

Bitumen is a naturally occurring material that is found in deposits where the permeability is low and passage of fluids through the deposit can only be achieved by prior application of fracturing techniques. Tar sand bitumen is a high-boiling material with little, if any, material boiling below 350°C (660°F); the boiling range is approximately the same as the boiling range of an atmospheric residuum.

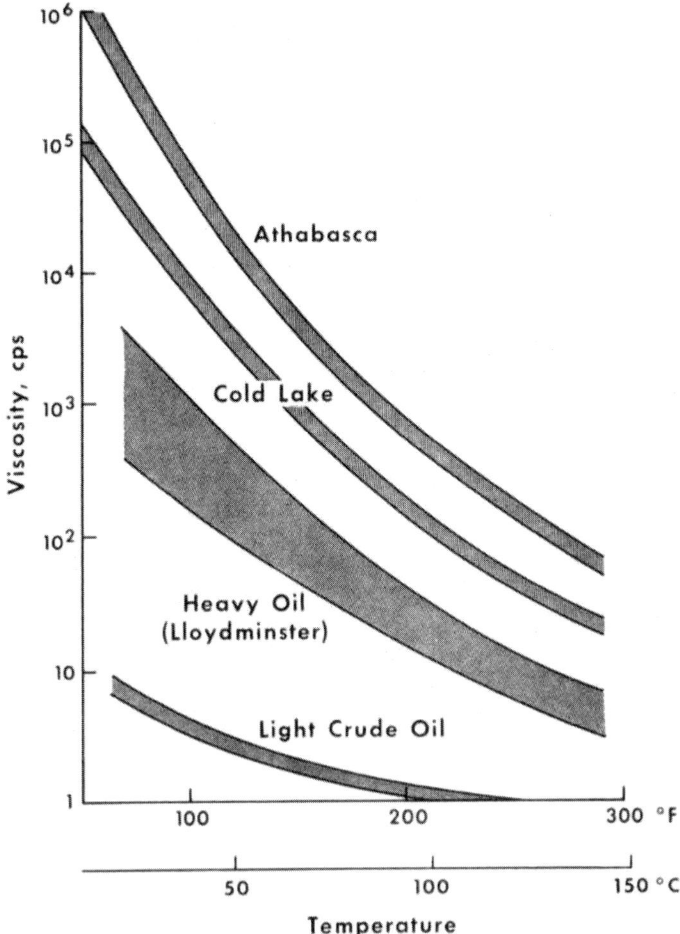

Figure 1–7 *General relationship of viscosity to temperature.*

Tar sands have been defined by the United States Federal Energy Administration (FE-76-4) as

> *...the several rock types that contain an extremely viscous hydrocarbon which is not recoverable in its natural state by conventional oil well production methods including currently used enhanced recovery techniques. The hydrocarbon-bearing rocks are variously known as bitumen-rocks oil, impregnated rocks, oil sands, and rock asphalt.*

The recovery of the bitumen depends to a large degree on the composition and construction of the sands. Generally, the bitumen found in tar sand deposits is an extremely viscous material that is *immobile under reservoir conditions* and cannot be recovered through a well by the application of secondary or enhanced recovery techniques.

The expression tar sand is commonly used in the petroleum industry to describe sandstone reservoirs that are impregnated with a heavy, viscous black crude oil that cannot be retrieved through a well by conventional production techniques (FE-76-4). However, the term tar sand is actually a misnomer; more correctly, the name *tar* is usually applied to the heavy product remaining after the destructive distillation of coal or other organic matter (Speight, 1994). Current recovery operations of bitumen in tar sand formations are predominantly focused on a mining technique.

It is incorrect to refer to native bituminous materials as tar or *pitch*. Although the word tar is descriptive of the black, heavy bituminous material, it is best to avoid its use with respect to natural materials and to restrict its meaning to the volatile or near-volatile products produced in the destructive distillation of such organic substances as coal (Speight, 1994). In the simplest sense, pitch is the distillation residue of the various types of tar.

Thus, alternative names, such as *bituminous sand* or *oil sand*, are gradually finding usage, with the name bituminous sands being more technically correct. The term oil sand is also used in the same way as the term tar sand; these terms are used interchangeably throughout this text.

Bituminous rock and bituminous sand are those formations in which the bituminous material is found filling in veins and fissures in fractured rocks or impregnating relatively shallow sand, sandstone, and limestone strata. This is, in fact, one geologically correct description of tar sand. The deposits contain as much as 20% bituminous material, and if the organic material in the rock matrix is bitumen, it is usual (although chemically incorrect) to refer to the deposit as rock asphalt to distinguish it from bitumen that is relatively mineral free. A standard test (ASTM D4) is available for determining the bitumen content of various mixtures with inorganic materials, although the use of the word bitumen as applied in this test might be questioned and it might be more appropriate to use the term *organic residues* to include tar and pitch. If the material is of the asphaltite-type or asphaltoid-type, the corresponding terms should be used: rock asphaltite or rock asphaltoid.

Bituminous rocks generally have a coarse, porous structure, with the bituminous material in the voids. It is a very common situation for the organic material to be present as an inherent part of the rock composition insofar as it is a diagenetic residue of the organic material detritus that was deposited with the sediment. The organic components of such rocks are usually refractory and are only slightly affected by most organic solvents.

A special class of bituminous rocks that has achieved some importance is the so-called *oil shale*. These are argillaceous, laminated sediments of generally high organic content that can be thermally decomposed to yield appreciable amounts of oil, commonly referred to as *shale oil*. Oil shale does not yield shale oil without the application of high temperatures and the ensuing thermal decomposition that is necessary to decompose the organic material (*kerogen*) in the shale.

Sapropel is an unconsolidated sedimentary deposit rich in bituminous substances. It is distinguished from peat by being rich in fatty and waxy substances and poor in cellulosic material. When consolidated into rock, sapropel becomes oil shale, bituminous shale, or boghead coal. The principal components are certain types of algae that are rich in fats and waxes. Minor constituents are mineral grains and decomposed fragments of spores, fungi, and bacteria. The organic materials accumulate in water under reducing conditions.

1.5 Validity of the Definitions

The validity of the definitions related to petroleum, heavy oil, and tar sand bitumen is subject to much scrutiny and, consequently, criticism.

Thus, although a single property number, such as API gravity or viscosity, is employed for some of the definitions, the validity of using a single property number is open to serious error since the number is subject to the experimental error or experimental differences of the analytical method by which the number was determined. Comparative properties, such as pour point and reservoir temperature, offer some logic for understanding the differences in behavior of heavy oil and tar sand bitumen.

For example, the generic term heavy oil is often applied to petroleum that has an API gravity of less than 20° and the term bitumen applied to those materials having less than 10° API. Following from this convenient generalization, there has also been an attempt to classify petroleum, heavy oil, and tar sand bitumen using viscosity scale, with 10,000 centipoise being the fine line of demarcation between heavy oil and tar sand bitumen. Use of such a system leads to confusion when having to differentiate between a material having a viscosity of 9,950 centipoise and one having a viscosity of 10,050 centipoise, taking into account the limits of accuracy of the method of viscosity determination. Whether the limits are the usual laboratory experimental difference (±3%) or more likely the limits of accuracy of the method (±5% to ±10%), there is the question of accuracy when tax credits for recovery of heavy oil and bitumen are awarded. In fact, the inaccuracies (i.e., the limits of experimental difference) of the method of measuring viscosity (or any single property) also increase the potential for misclassification using this single property for classification purposes.

Any attempt to classify petroleum, heavy oil, and bitumen on the basis of a single property is no longer sufficient to define the nature and properties of petroleum and petroleum-related materials. The general classification of petroleum into conventional petroleum, heavy oil, and extra heavy oil should involve not only an inspection of several properties but also some acknowledgment of the method of recovery.

Petroleum is referred to generically as a *fossil energy resource* and is further classified as a *hydrocarbon resource;* for illustrative (or comparative) purposes in this text, coal and oil shale kerogen have also been included in this classification. However, the inclusion of coal and oil shale under the broad classification of hydrocarbon resources has required (incorrectly) that the term *hydrocarbon* be expanded to include the high molecular weight (macromolecular) non-hydrocarbon heteroatomic species that constitute coal and oil shale kerogen. Heteroatomic species are those organic constituents that contain atoms other than carbon and hydrogen, e.g. nitrogen, oxygen, sulfur, and metals (nickel and vanadium), as an integral part of the molecular matrix.

Use of the term organic sediments is more correct and to be preferred (Figure 1–8). The inclusion of coal and oil shale kerogen in the category hydrocarbon resources is due to the fact that these two natural resources (coal and oil shale kerogen) will produce hydrocarbons on high-temperature processing (Figure 1–9). Therefore, if either coal

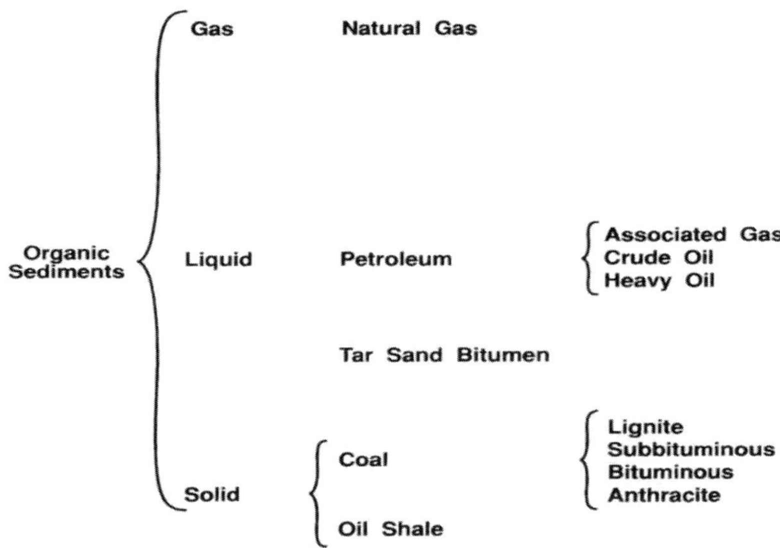

Figure 1–8 *Classification of fossil fuel as organic sediments.*

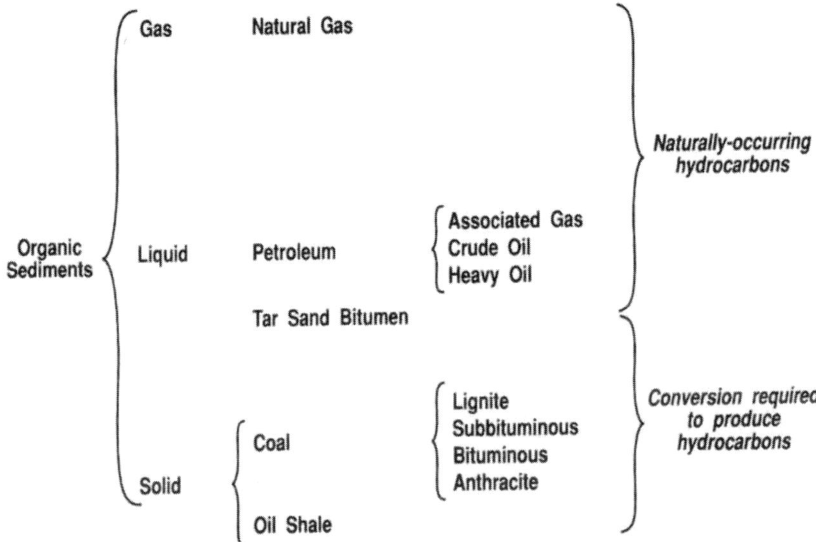

Figure 1–9 *Classification of fossil fuels as hydrocarbon resources and hydrocarbon producing resources.*

and/or oil shale kerogen is to be included in the term hydrocarbon resources, it is more appropriate that they be classed as *hydrocarbon-producing resources* under the general classification of *organic sediments*. Thus, fossil energy resources divide into two classes: (1) naturally occurring hydrocarbons (petroleum, natural gas, and natural waxes) and (2) hydrocarbon sources (oil shale and coal), which may be made to generate hydrocarbons by the application of conversion processes. Both classes may very aptly be described as organic sediments.

Whenever attempting to define or classify tar sand bitumen, it is always necessary to return to the definition as given by the United States Federal Energy Administration (FE-76-4) (above). By inference, petroleum and heavy oil are recoverable by well production methods and currently used enhanced recovery techniques. For convenience, it is assumed that before depletion of the reservoir energy, conventional crude oil is produced by primary and secondary techniques whereas heavy oil requires tertiary (enhanced) oil recovery (EOR) techniques. While this is an oversimplification, it may be used as a general guide.

The term *natural state* cannot be defined out of context; in the context of FEA Ruling 1976-4, the term is defined in regards to the composition of the heavy oil or bitumen. The final determinant of whether or not a reservoir is a tar sand deposit is the character of the viscous phase (bitumen) and the method that is required for recovery.

Generally, bitumen is solid or near solid at room temperature and is solid or near solid at reservoir temperature. In other words, tar sand bitumen is immobile in the reservoir and requires conversion or extreme stimulation for recovery.

Thus, by this definition (FE-76-4), tar sand bitumen is not crude oil, and it is set apart from conventional crude oil and heavy crude oil insofar as it cannot be recovered from a deposit by the use of conventional (including enhanced) oil recovery techniques as set forth in the June 1979 Federal Energy Regulations. To emphasize this point, bitumen has been recovered commercially by mining and the hot water process and is currently upgraded (converted to synthetic crude oil) by a thermal or hydrothermal process followed by product hydrotreating to produce a low-sulfur hydrocarbon product known as *synthetic crude oil*.

Tar sand bitumen is a naturally occurring material that is immobile in the deposit and cannot be recovered by the application of enhanced oil recovery technologies, including steam-based technologies. On the other hand, heavy oil is mobile in the reservoir and can be recovered by the application of enhanced oil recovery technologies, including steam-based technologies.

Since the most significant property of tar sand bitumen is its *immobility* under the conditions of temperature and pressure in the deposit, the interrelated properties of API gravity (ASTM D287) and viscosity (ASTM D445) may present an *indication* (but only an indication) of the mobility of oil or immobility of bitumen. In reality, these properties only offer subjective descriptions of the oil in the reservoir. The most pertinent and objective representation of this oil or bitumen mobility is the *pour point* (ASTM D97) (see also Chapter 4).

By definition, the pour point is the lowest temperature at which oil will move, pour, or flow when it is chilled without disturbance under definite conditions (ASTM D97). In fact, the pour point of an oil when used in conjunction with the reservoir temperature gives a better indication of the condition of the oil in the reservoir than the viscosity. Thus, the pour point and reservoir temperature present a more accurate assessment of the condition of the oil in the reservoir, being indicators of the mobility of the oil in the reservoir. When used in conjunction with reservoir temperature, the pour point gives an indication of the liquidity of the heavy oil or bitumen and, therefore, the ability of the heavy oil or bitumen to flow under reservoir conditions. In summary, the pour point is an important consideration because, for efficient production, additional energy must be supplied to the reservoir by a thermal process to increase the reservoir temperature beyond the pour point.

For example, Athabasca bitumen with a pour point in the range 50 to 100°C (122 to 212°F) and a deposit temperature of 4 to 10°C (39 to 50°F) is a solid or near solid in the deposit and will exhibit little or no mobility under deposit conditions. Pour points of 35 to 60°C (95 to 140°F) have been recorded for the bitumen in Utah, with formation temperatures on the order of 10°C (50°F). This indicates that the bitumen is solid within the deposit and therefore immobile. The injection of steam to raise and maintain the reservoir temperature above the pour point of the bitumen and to enhance bitumen mobility is difficult, in some cases almost impossible. Conversely, when the reservoir temperature exceeds the pour point, the oil is fluid in the reservoir and therefore mobile. The injection of steam to raise and maintain the reservoir temperature above the pour point of the bitumen and to enhance bitumen mobility is possible and oil recovery can be achieved.

A method that uses the pour point of the oil and the reservoir temperature (Figure 1–10) adds a specific qualification to the term *extremely viscous* as it occurs in the definition of tar sand. In fact, when used in conjunction with the recovery method (Figure 1–11), pour point offers more general applicability to the conditions of the oil in the reservoir or the bitumen in the deposit, and comparison of the two temperatures (pour point and reservoir temperatures) shows promise and may find more general use.

1.6 Conclusions

In summary, heavy oil is more viscous than conventional petroleum, but this resource is in plentiful supply and different methods of production are required.

Heavy oil cannot be defined adequately or with any degree of accuracy using a single property. Likewise, tar sand bitumen cannot be defined using a single property. Both, however, can be redefined by the recovery method. By inference, heavy oil can also be defined using the same definition as tar sand bitumen. Heavy oil is usually mobile in the reservoir, whereas tar sand bitumen is immobile in the deposit.

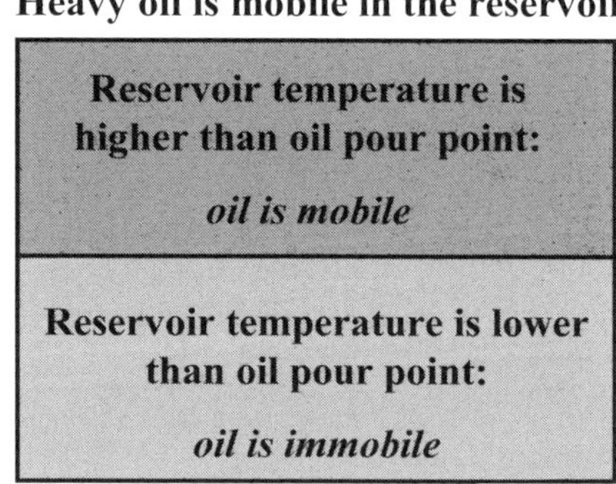

Figure 1–10 *Simplified of the use of pour point to define heavy oil and bitumen.*

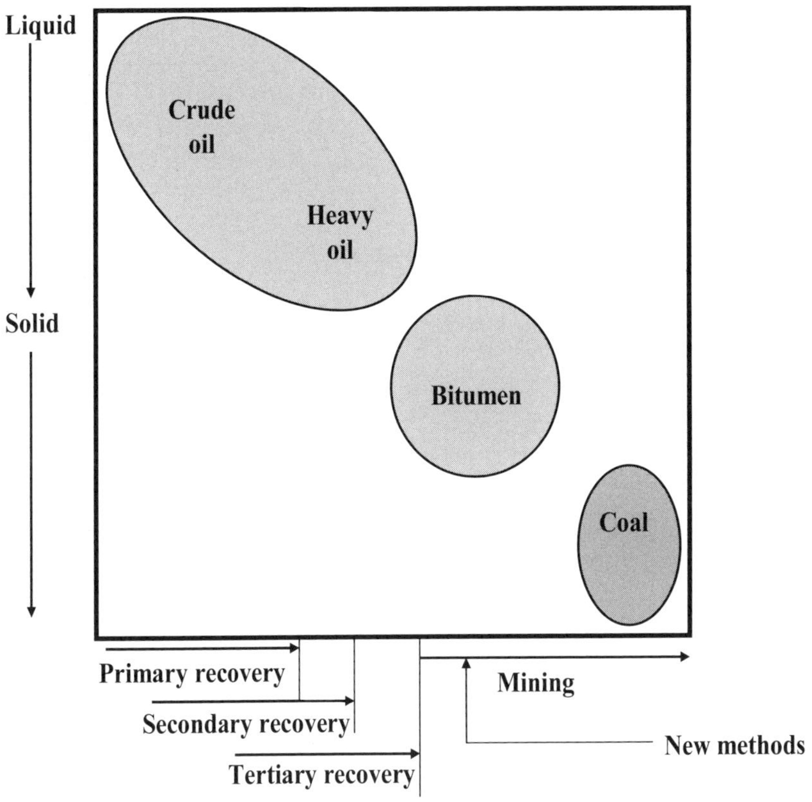

Liquid
Solid
Crude oil
Heavy oil
Bitumen
Coal
Primary recovery
Secondary recovery
Tertiary recovery
Mining
New methods

Figure 1–11 *Schematic representation of the properties and recovery methods for crude oil, heavy oil, bitumen, and coal.*

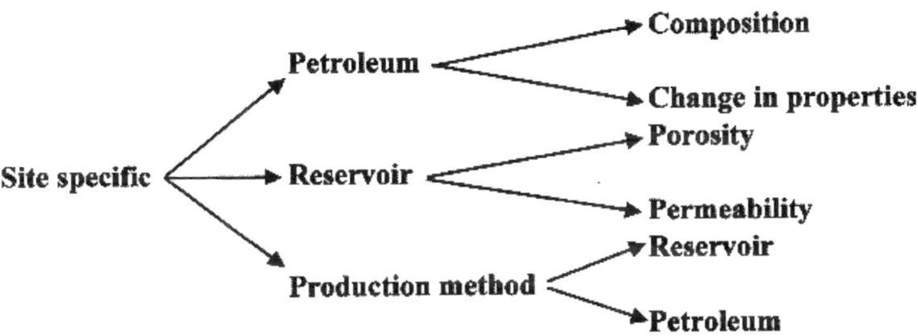

Figure 1–12 *Representation of the changing parameters for crude oil and/or heavy oil.*

Finally, it is essential to realize that in the current context of conventional petroleum and heavy oil, there are several parameters that can influence properties and recovery. These properties are usually site specific to the particular reservoir in which the crude oil or heavy oil is located (Figure 1–12).

1.7 References

Abraham, H. 1945. Asphalts and Allied Substances. Van Nostrand, New York.

Ancheyta, J., and Speight, J.G. 2007. Hydroprocessing of Heavy Oils and Residua. CRC-Taylor and Francis Group, Boca Raton, Florida.

ASTM D4. Standard Test Method for Bitumen Content. Annual Book of Standards. American Society for Testing and Materials, West Conshohocken, Pennsylvania.

ASTM D97. Standard Test Method for Pour Point of Petroleum Products. Annual Book of Standards. American Society for Testing and Materials, West Conshohocken, Pennsylvania.

ASTM D287. Standard Test Method for API Gravity of Crude Petroleum and Petroleum Products (Hydrometer Method). Annual Book of Standards. American Society for Testing and Materials, West Conshohocken, Pennsylvania.

ASTM D445. Standard Test Method for Kinematic Viscosity of Transparent and Opaque Liquids (and Calculation of Dynamic Viscosity). Annual Book of Standards. American Society for Testing and Materials, West Conshohocken, Pennsylvania.

ASTM D4175. Standard Terminology Relating to Petroleum, Petroleum Products, and Lubricants. Annual Book of Standards. American Society for Testing and Materials, West Conshohocken, Pennsylvania.

British Petroleum Company. 2007. Statistical Review of World Energy. British Petroleum Company, London,.

Cobb, C., and Goldwhite, H. 1995. Creations of Fire

Chemistry's Lively History from Alchemy to the Atomic Age. Plenum Press, New York.

Forbes, R. J. 1958a. A History of Technology. Oxford University Press, Oxford, England.

Forbes, R. J. 1958b. Studies in Early Petroleum Chemistry. E. J. Brill, Leiden, The Netherlands.

Forbes, R.J. 1959. More Studies in Early Petroleum Chemistry. E.J. Brill, Leiden, The Netherlands.

Forbes, R. J. 1964. Studies in Ancient Technology. E. J. Brill, Leiden, The Netherlands.

International Energy Agency. 2005. Resources to Reserves: Oil & Gas Technologies for the Energy Markets of the Future. International Energy Agency, Paris, France.

Speight, J.G. 1978. Personal observations at archeological digs at the cities of Babylon, Calah, Nineveh, and Ur.

Speight. J. G. 1984. Characterization of Heavy Crude Oils and Petroleum Residues. Ed. S. Kaliaguine and A. Mahay. Elsevier, Amsterdam. p. 515.

Speight, J. G. 1994. The Chemistry and Technology of Coal. 2nd Edition. Marcel Dekker. New York.

Speight. J. G. 2000. The Desulfurization of Heavy Oils and Residua. 2nd Edition. Marcel Dekker, New York.

Speight, J.G., and Ozum, B. 2002. Petroleum Refining Processes. Marcel Dekker Inc., New York.

Speight, J.G. 2007. The Chemistry and Technology of Petroleum. 4th Edition. CRC-Taylor and Francis Group, Boca Raton, Florida.

Taylor, S. 2006. Keys to Heavy Oil: Characterizing Fluids and the Reservoir. Schlumberger.

Totten, G.E. 2007. A Timeline of Highlights from the Histories of ASTM D2 Committee and the Petroleum Industry.

http://www.astm.org/COMMIT/D02/to1899_index.html.

CHAPTER 2

ORIGIN AND OCCURRENCE

The declining reserves of light crude oil have resulted in an increasing need to develop options to upgrade the abundant supply of known heavy oil reserves (IEA, 2005; Meyer and Attanasi, 2003; Meyer et al., 2007). In addition, there is considerable focus and renewed efforts on adapting recovery techniques to the production of heavy oil.

Over the past decade, the demand for crude oil worldwide has substantially increased, straining the supply of conventional oil. This has led to consideration of alternative or insufficiently utilized energy sources, especially heavy crude oil to supplement short- and long-term needs. Heavy oil has been used as refinery feedstock for considerable time, usually blending with more conventional feedstocks, but has commanded lower prices because of its lower quality relative to conventional oil.

Obviously, differences exist between heavy oil and conventional oil, according to the volatilities of the constituents. When the lower-boiling constituents are lost through natural processes after evolution from organic source materials, the oil becomes heavy, with a high proportion of asphaltic molecules and with substitution in the carbon network of heteroatoms such as nitrogen, sulfur, and oxygen. Therefore, heavy oil, regardless of source, always contains the heavy fractions, the asphaltic materials, which consist of resins and asphaltenes (Figure 2–1). Removal or reduction of the asphaltene fraction, through deasphalting or leaving these constituents in the reservoir during recovery, improves the refinability of heavy oil.

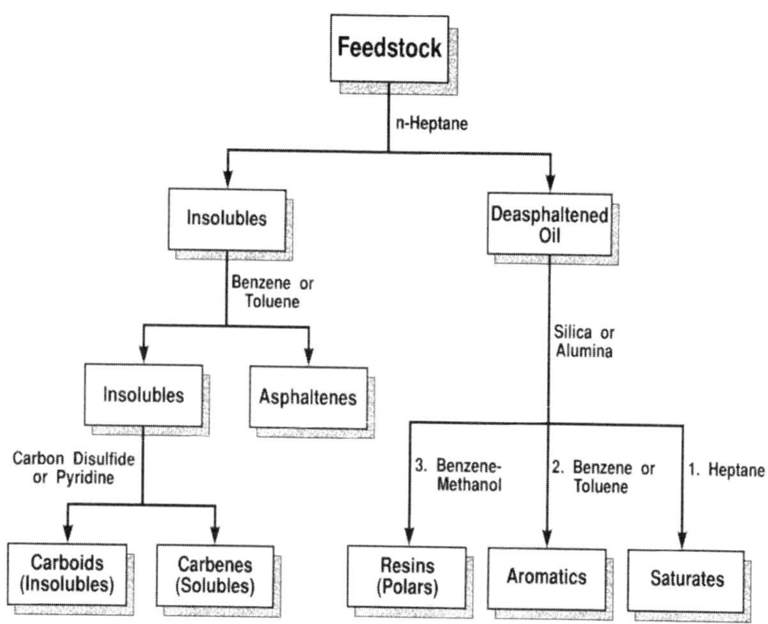

Figure 2–1 *Separation scheme and nomenclature of different fractions of petroleum and heavy oil—the nomenclature is an artifact of the separation process and does not guarantee chemical types or chemical homogeneity.*

The significance of the absence of the asphaltene constituents is reflected in the capital and operating expenses required for the recovery, transportation, refining, and environmental mitigation.

Detection of hydrocarbons in the subsurface during exploration takes a number of forms: direct identification of hydrocarbons at the surface, direct identification of hydrocarbon indicators (DHI) in the subsurface, and indirect identification of indicators both at the surface and in the subsurface. Traditionally, oil exploration was primarily conducted by recognizing seeps of hydrocarbons at the surface. The Chinese, for example, used oil (mostly bitumen) obtained from seeps for use in medication, waterproofing, and warfare several thousand years ago. The ancient Chinese frequently dug shallow pits or horizontal tunnels at the seep locations in order to recover the oil. In Baku, Azerbaijan, there are still gas and oil seeps that are permanently alight and have been used to light caravanserai since the times of Marco Polo and the Silk Road. With the dawning of the modern era in Oil Creek, Pennsylvania, Colonel Edwin Drake drilled the first well to intentionally look for oil in the subsurface in 1859. Again, this was based on direct identification of seeped hydrocarbons at the surface. Initially, the oil was used to provide kerosene for lamps, but the later invention of automobiles drove up demand and ushered in modern methods of oil exploration.

Around the turn of the century and up until the 1950s, the main exploration tool used for finding oil was the use of intensive and detailed geological mapping. This was frequently in terrain that was remote and inhospitable. The early pioneers working their way through the jungles of Burma, the deserts of Iraq, or the mountains of Iran would conduct detailed evaluations of the nature and distribution of rock units that could represent potential reservoirs, seals, and source units, as well as the frequency, orientation, and geological history of folds or faults that could act as traps for the migrating hydrocarbons. If all four of the features required for oil or gas to be created and trapped could be recognized in a region, then a variety of play concepts could be generated. Detailed local study might identify a suitable target (prospect), and then a shallow well would be drilled to test the features.

One of the most important recent discoveries in petroleum studies has been plate tectonics. Not only has plate tectonics revolutionized the earth sciences, but they also have provided a conceptual setting for oil exploration. The movement of plates around the surface of Earth creates large-scale depressions into which substantial quantities of sediments eroded from the surrounding high ground may accumulate. These accumulations can exceed thicknesses of several thousand kilometers and are referred to as sedimentary basins. By comparing basins around the world and by analogy to existing producing hydrocarbon regions, an exploration team can say which basins are worth looking at in more detail. Then the exploration team spends time ensuring that within such a basin there exist all the key elements that control the presence of hydrocarbons. Assuming that all the needed features are present, the team agrees that the basin contains a viable petroleum system, and prospect generation can proceed.

In modern exploration programs, the mapping of gravity and magnetic anomalies would normally be the first two methods to be applied to a new basin or region being evaluated. These techniques would be used to identify large-scale changes in the structure of the basement and sedimentary basins, and major differences in rock density, such as the influx of dense igneous rocks or light salt into a sedimentary sequence. These techniques are large scale, can be applied over both land and water, and can even be collected remotely from plane or satellite.

At the same time, remote sensing of onshore areas by large-scale photogeological surveys and, after the 1970s, by satellite imaging, can identify areas with anticlinal and faulted structural features, seeps, or salt domes frequently associated with oil occurrences. Offshore remote sensing of the sea surface can lead to the identification of slicks associated with the seepage of oil (both natural and manmade) into the water column. A coarse two-dimensional grid of seismic data is then collected to obtain a picture of the subsurface in the area to be targeted. Seismic data collection involves the generation of a seismic wave using an energy source such as an air-gun in water, dynamite in drill holes inland, or a truck with a plate that is thumped down onto the road/soil surface (vibroseis). The wave travels through the earth's rock layers and reflects back off key surfaces. The time it takes for the waves to be received back at the surface along with the waves' strength is recorded via geophones and displayed on a seismic section. Processing the two-dimensional seismic sections using highly sophisticated software reveals the detailed structure of the subsurface and in certain circumstances shows the presence of direct hydrocarbon indicators such as bright spots associated with gas/water differences. Primarily, though, seismic data are used to indicate the nature of folded and faulted structures that could prove to be suitable hydrocarbon traps. These structures are frequently referred to as leads.

The objective of seismic acquisition and processing is to acoustically image the subsurface in a geologically accurate manner with the highest resolution possible. For a detailed analysis of a small area representing a field or prospect, a high-density and calibrated three-dimensional seismic survey is performed and the data are collected. Modern technology also allows scientists to accurately map changes in fluid movements through time (repeat multiple 3-D seismic surveys, known as 4-D seismic survey). This technique is now particularly important in monitoring production performance of the reservoir.

Ultimately, however, the only way of confirming the presence or absence of hydrocarbons at depth is by drilling the prospect. In certain areas of the world where drilling is cheap and the subsurface has been explored extensively, such as certain onshore basins of the United States, drilling is commonly preferred to extensive and expensive seismic acquisition. Wells are then analyzed using electric, sonic, and radioactive logging techniques that measure characteristics of the rocks and fluids. These methods can identify the presence of oil and gas, which can then be tested to see if they occur at commercially viable production levels. On the other hand, at a cost of over 10 million dollars per offshore exploration well, the oil companies are likely to employ the sophisticated battery of direct and indirect detection techniques before resorting to drilling in these areas.

2.1 Origin of Petroleum and Heavy Oil

There are two theories on the origin of carbon fuels: the abiogenic theory and the biogenic theory (Kenney et al., 2001). The two theories have been intensely debated since the 1860s, starting shortly after the discovery of the widespread occurrence of petroleum. It is not the intent of this section to sway the reader in his or her views of the origin of petroleum and natural gas. The intent is to place before the reader both points of view; the reader can research further to decide.

In general, heavy oil was originally conventional oil that migrated from deep source rocks or deep reservoirs to near the surface, where the oil was biologically degraded and weathered by water. Bacteria feeding on the migrated conventional oil removed hydrogen and produced the denser, more viscous heavy oil. Lighter hydrocarbons may also have evaporated from the shallow, uncapped formations.

Therefore, the origins of heavy oil are the same as the origins of conventional oil. A brief discussion of the means by which oil is formed is warranted here as a point of reference for heavy oil properties and behavior.

2.1.1 Abiogenic Origin

There have been several attempts at formulating theories that describe the detail of the origin of petroleum. The early postulates started with inorganic substances as source material. For example, in 1866, Berthelot considered acetylene the basic material and crude oil constituents as being produced from the acetylene.

$$CaCO_3 + \text{alkali metal} \rightarrow CaC_2 \text{ (calcium carbide)}$$

$$CaC_2 + H_2O \rightarrow HC{\equiv}CH \text{ (acetylene)} \rightarrow \text{petroleum}$$

There are also several more recent theories related to the formation of petroleum from non-biogenic sources in the earth (Gold, 1984, 1985; Gold and Soter, 1980, 1982, 1986; Osborne, 1986; Szatmari, 1989).

The idea of an abiogenic petroleum origin proposes that large amounts of carbon exist naturally, some in the form of hydrocarbons. Hydrocarbons are less dense than aqueous pore fluids, and they migrate upward through deep fracture networks. Thermophilic, rock-dwelling microbial life forms are in part responsible for the biomarkers found in petroleum. However, their role in the formation, alteration, or contamination of the various hydrocarbon deposits is not yet understood. Thermodynamic calculations and experimental studies confirm that n-alkanes (common petroleum components) do not spontaneously evolve from methane at pressures typically found in sedimentary basins, and so the theory of an abiogenic origin of hydrocarbons suggests deep generation (below 120 miles).

From the chemical point of view, the inorganic theories are interesting because of their historical importance, but these theories have not received much attention. Geological and chemical methods have demonstrated the optical activity of petroleum constituents, the presence of thermally labile organic compounds, and the almost exclusive occurrence of oil in sedimentary rocks.

2.1.2 Biogenic Origin

It is now generally accepted, but not conclusively proven, that petroleum formation predominantly arises from the decay of organic matter in the earth. Nevertheless, alternative theories should not be dismissed until it can be conclusively established that petroleum formation is due to one particular aspect of geochemistry.

It is generally proposed that the formation of petroleum constituents occurs through the progressive chemical change of materials provided by microscopic aquatic organisms that were incorporated over eons in marine or near-marine sedimentary rocks. In fact, the details of petroleum genesis (*diagenesis*, *catagenesis*, and *metagenesis*) have long been a topic of interest. However, the details of this *transformation* and the mechanism by which petroleum is expelled from the source sediment and accumulates in the reservoir rock are still uncertain.

Transformation of some of this sedimentary material to petroleum probably began soon after deposition, with bacteria playing a role in the initial stages and clay particles serving as catalysts. Heat within the strata may have provided energy for the reaction, temperatures increasing more or less directly with depth. Some evidence indicates that most petroleum has formed at temperatures not exceeding about 100 to 120°C (210 to 250°F), with the generation of petroleum hydrocarbons beginning at temperatures as low as 65°C (150°F).

Thus, it is possible for heavy oil to form by several processes. The oil may be expelled from its source rock as immature oil. There is general agreement that immature oils account for a small percentage of the heavy oil (Larter et al., 2006). Most heavy oil and natural bitumen is thought to be expelled from source rocks as light or medium oil, which subsequently migrates to a trap. If the trap is later elevated into an oxidizing zone, several processes can convert the oil to heavy oil. These processes include water washing, bacterial degradation (aerobic biodegradation), and evaporation. A third proposal is that biodegradation can also occur at depth in subsurface reservoirs (Head et al., 2003; Larter et al., 2003; Larter et al., 2006). This explanation permits biodegradation to occur in any reservoir that has a water leg and has not been heated to more than 80°C (176°F). The controls on the biodegradation depend on local factors rather than basinwide factors.

In addition, like its conventional- or light-oil counterpart, the composition of heavy oil is greatly influenced not only by the nature of the precursors that eventually form the heavy oil, but also by the relative amounts of these precursors (which are dependent upon the local flora and fauna) that occur in the source material. Hence, it is not surprising that heavy oil, like conventional petroleum, can vary in composition with the location and age of the reservoir. The lower mobility of heavy oil also makes it extremely likely that two wells in the same reservoir will produce heavy oil with different characteristics.

2.1.3 Occurrence and Distribution

Petroleum is found in sedimentary rocks throughout the world. In many places the oil has been degraded, and the result is heavy oil or tar sand bitumen, depending upon the degree of degradation (Meyer and Attanasi, 2004). Four main issues control the occurrence and distribution of conventional oil and heavy oil: (1) source, (2) reservoir, (3) seal, and (4) trap.

A source is a fine-grained rock unit containing sufficient organic matter so that when it is heated and/or placed under pressure (maturation), hydrocarbons are generated. Source rocks with organic matter of marine algal origin are most likely to generate oil under optimum maturation conditions. Rocks dominated by land plant matter, on the other hand, will tend to create gaseous hydrocarbons. The hydrocarbons are of lower gravity than the surrounding groundwater and, therefore, move away from and generally upwards (migrate) from the source rock until they are trapped in a reservoir.

Geologists generally agree that petroleum deposits formed from the remains of enormous quantities of aquatic plants and organisms that became mixed with sand and mud at the bottom of bodies of water. If a certain regime of temperatures and pressures existed, this biological material was converted into petroleum during geologic time. The pressure in the earth's crust forced the petroleum into the tiny open spaces between the grains of imbedded sandstone and other coarse textured strata (and not into vast underground pools).

A reservoir is a rock unit that acts as a storage device for the hydrocarbons that migrate from the source rock. Hydrocarbons are retained within the reservoir because these rocks contain numerous pores (essentially microscopic holes) between the mineral grains making up the fabric of the reservoir. In good-quality reservoirs, the porosity is frequently over 20% of the rock volume. However, the pores need to be interlinked in such a manner that the fluids can move into (and out of, if we are to exploit the oil and gas) the reservoirs over geological time. This is known as permeability. There are two main rock types that make up the giant reservoirs around the world—sandstones, which are made up of sand grains (quartz and feldspar in the majority), and carbonates, which are made up of organically created calcium carbonate grains (corals, algae, and shells) or mud. In order to stop the upward movement of hydrocarbons and constrain them to one zone of the subsurface (trap), there must be a barrier to prevent fluid migration. Generally, this mechanism or seal consists of rocks that are impermeable to fluid flow. The most effective of these seals are mudstone or shale, very fine-grained rocks containing abundant clay minerals. Occasionally, the impermeable layers are dense igneous rocks, and in rare situations, there may be significant rock and fluid pressure differences in a region that prevent fluid flow and acts as a seal.

Equally important is the presence of a trapping mechanism. For petroleum to be present in commercial quantities, some sort of trapping mechanism must have been present to prevent its escape. In addition, these traps must have not been breached by natural means after accumulation occurred. There are two major kinds of trapping mechanisms. Structural traps are created by a deformation of the earth's crust; the folding or faulting of rocks results in the entrapment of petroleum. Stratigraphic traps result from the relative differences in the porosity and permeability of the oil-bearing rocks compared to less porous and permeable adjacent rocks, which serve to prevent the further movement of oil. Many traps have structural and stratigraphic features.

Since oil reservoirs (rock formations or traps holding an accumulation of petroleum) were formed from sea sediments, most reservoirs also contain salt water. In addition, natural gas is almost always present, either dissolved in the oil or as free gas separate from the oil. The associated water and natural gas are important in maintaining pressure during petroleum production.

The biogenic theory of petroleum formation and occurrence has led to the search for thick beds of sedimentary rocks. Geophysical techniques have aided in the discovery of approximately six hundred basins (tracts of land in which the rock strata are tilted toward a common center), both onshore and offshore, which may contain oil or natural gas. Sufficient seismic work has been done to give an indication of their prospective petroleum area and their general structural aspects. In approximately four hundred of these basins, exploratory drilling in varying amounts and degrees of success has taken place. Commercial quantities of oil and gas are being produced from approximately 160 of these basins.

There remain approximately 200 basins which have not been drilled. This lack of drilling has been due to many factors, including the location of the areas, restriction on access caused by individual governments and territorial disputes, and most importantly, judgments about their potential for yielding petroleum. In particular, some of these basins lie in the offshore Arctic; past prices of oil have not been sufficient to justify exploration there (although this appears to be changing).

Although oil has been found in commercial quantities in approximately 160 basins, 25 of the 160 basins (containing discoveries of over 10 billion barrels) have accounted for over 80% of total discoveries. Not only have oil discoveries been concentrated in a limited number of basins, but the majority of oil has been found in a relatively small number of large fields. Over 90% of the oil has been found in a small number of large fields containing at least 100 million barrels of liquid petroleum or liquid-equivalent (liquids plus natural gas) resources. Of the more than 20,000 fields discovered, 1700 contain over approximately 900 of the 1000 billion barrels which had been produced or were known to exist at the end of 1975. Further, an even smaller number of approximately 280 giant fields (containing at least 500 million barrels in liquids or liquid equivalents) contained approximately 75% of the oil found.

In the U.S., a smaller share of oil has been found in large fields than the worldwide average of 90%. However, approximately 70% of the oil found in the U.S. has been found in large fields containing at least 100 million barrels of liquid petroleum or liquid equivalent. The significance of large fields for the U.S. is illustrated by the petroleum discoveries in the Permian Basin, which were responsible for 18% of the crude oil and 10% of the natural gas discovered in the U.S. through 1974. The Permian Basin has been extensively explored, resulting in the discovery of over 4000 oil and gas fields by the end of 1974. At that time, over 60% of the oil and gas had been discovered in the 70 largest fields. whereas only 2% had been discovered in the 2700 smallest fields (fields containing less than 1 million barrels of oil or equivalent).

Many experts expect the importance of large fields for new discoveries to continue, for both geologic and economic reasons. In new offshore areas, only large fields are economical to develop at today's oil prices ($30 per barrel), although in places where the associated infrastructure (pipelines, etc.) exists, smaller finds may be sought and developed. The earliest discoveries of oil and gas made over a century ago were based on surface seepages and analyses of surface geology. After a century of experience, petroleum exploration has become extremely sophisticated. New techniques and instrumentation, such as geophysical surveys, geochemical analysis, and subsurface logging, combined with the use of computers, provide information on structural characteristics of the subsurface.

These techniques provide only raw data which must be evaluated by geologists and geophysicists; the room for judgment and, therefore, differences in interpretation remains great. Drilling remains the only way to determine the actual existence of petroleum resources, with the other exploration techniques providing guidance as to the most promising locations within a basin.

An example illustrating the geologic uncertainties and risks in oil exploration is the Destin Dome located in the Gulf of Mexico. Over 600 million dollars were paid by oil companies for leases around the Destin Dome in view of its extremely promising petroleum potential. After drilling many dry holes, the companies involved returned the leases to the government with no petroleum produced.

2.2 Reservoirs

Heavy oils typically occur in geologically young reservoirs (from the Cretaceous) (Table 2–1). Because these reservoirs are shallow, they have less effective seals and are thus exposed to conditions conducive to the formation of heavy oils. The fact that most heavy oil reservoirs are shallow is the reason why many of them were discovered as soon as human beings settled nearby. Collecting oil from seeps and digging by hand were the earliest and most primitive means of recovery, followed by mining and tunneling.

However, heavy oil resources, along with tar sand bitumen resources, can be subdivided into a number of different categories based on their location, environment, and characteristics. The following categorization is not all-encompassing, but it does illustrate the wide variety among heavy oil resources:

- Shallowest resources (<150 feet depth)
- Shallow resources (150 to 300 feet deep, with no cap rock seal)
- Medium-depth resources (300 to 1,000 feet deep, cap rock seal, pressure <200 psi)
- Intermediate-depth resources (1,000 to 3,000 feet, pressure >200 psi)
- Deep resources (>3,000 feet deep)
- Carbonate resources (tight rock formations, variable porosity)
- Thinly bedded resources (<30 feet thick)
- Highly laminated resources (low vertical permeability, often shale layering).

Table 2–1 The Geologic Timescale

Era	Period	Epoch	Duration (millions of years)	Years ago (millions of years)
Cenozoic	Quaternary	Holocene	10,000 years ago to the present	
		Pleistocene	2	0.01
	Tertiary	Pliocene	11	2
		Miocene	12	13
		Oligocene	11	25
		Eocene	22	36
		Paleocene	71	58
Mesozoic	Cretaceous		71	65
	Jurassic		54	136
	Triassic		35	190
Paleozoic	Permian		55	225
	Carboniferous		65	280
	Devonian		60	345
	Silurian		20	405
	Ordovician		75	425
	Cambrian		100	500
Precambrian			3,380	600

Properties of the heavy oil, such as composition and viscosity, are equally important properties but have not been included in the above list. If the oil properties were to be included, the complexity of the categorization of the heavy oil resources could increase by at least an order of magnitude.

In addition, many heavy oil reservoirs have been found in Arctic regions and offshore beneath the continental shelves of Africa and

North and South America. Heavy oil has also been discovered beneath the Caspian, Mediterranean, Adriatic, Red, Black, North, Beaufort, and Caribbean Seas, as well as beneath the Persian Gulf and the Gulf of Mexico.

Most of the heavy oil currently recovered is produced from underground reservoirs. However, surface seepage of crude oil and natural gas are common in many regions. In fact, it is the surface seepage of oil that led to the first use of the high-boiling material (heavy oil or bitumen) in the Fertile Crescent. It may also be stated that the presence of active seeps in an area is evidence that oil and gas are still migrating.

Heavy oil reservoirs are usually shallow (up to 1000 meters below the surface line) and therefore present low reservoir temperatures (between 40 and 60°C). Low sedimentary overburden tends to ease the biodegradation of the oil, especially when associated to bottom aquifers. The shallow depth of these reservoirs also contributes to the formation of geomechanically fragile structures where faults create geological compartments and heterogeneities. This kind of reservoir may also have low seal pressure, which may cause the dissolved gas to leave the oil, increasing its viscosity. The reservoir lithology is usually sandstones deposited as turbidites; high permeability (in the order of Darcies) and porosity are quite common. High permeability may compensate for the elevated oil viscosity, resulting in high well productivities.

Although one of the characteristics of heavy oil is that it is mobile in the reservoir, i.e., the pour point of the oil is lower that the reservoir temperature (Chapter 1), some reservoirs have a sufficiently high temperature that heavy oil can be produced with essentially conventional methods. Once the oil is produced at the surface, the temperature differential between the reservoir (higher temperature) and the surface (lower temperature) might be such that the oil resorts to the more familiar extremely viscous fluid.

Finally, just as the properties of the oil can dictate the relative ease or difficulty of production, the same applies for the reservoir properties. Heavy oil usually requires specialized (tertiary, enhanced) production methods. Very shallow heavy oil reservoirs can be mined and the oil allowed to drain into a mine tunnel. Slightly deeper deposits can be produced by increasing reservoir contact with horizontal wells and multilaterals (producing the oil with large amounts of sand) or by injecting steam, which lowers the viscosity and reduces the residual oil saturation, thus improving recovery efficiency.

2.3 Reserves

Oil *reserves* are the estimated quantities of conventional petroleum and/or heavy crude oil that are claimed to be recoverable under existing economic and operating conditions. Many oil-producing nations do not reveal their reservoir engineering field data, and instead they provide unsubstantiated claims for their oil reserves.

In most cases, *oil* refers to conventional oil, but, depending on the source, tar sand bitumen may or may not be included. The exact definition varies from country to country, and national statistics are not always comparable and may even be manipulated for political reasons.

The total amount of conventional oil or heavy oil in a reservoir is known as *oil in place*. However, because of reservoir characteristics and limitations in production technology, only a fraction of this oil can be brought to the surface, and it is only this producible fraction that is considered to be reserves. The ratio of reserves to oil in place for a given field is often referred to as the *recovery factor*. The recovery factor of a field may change over time based on operating history and in response to changes in technology and economics. The recovery factor may also rise over time for conventional oil if additional investment is made in enhanced oil recovery techniques such as gas injection or waterflooding. For heavy oil, the evolution of older methods and the use of newer and innovative methods may increase the recovery factor. Because the geology of the subsurface cannot be examined directly, indirect techniques must be used to estimate the size and recoverability of a resource. While new technologies have increased the accuracy of these techniques, significant uncertainties still remain. In general, most early estimates of the reserves of an oil field are conservative and tend to grow with time; this phenomenon is referred to as *reserves growth* (Morehouse, 1997).

2.3.1 Definitions

The definitions that are used to describe petroleum and heavy oil reserves are often misunderstood because these terms are not adequately defined at the time of use (Speight, 2007). Therefore, as a means of alleviating this problem, it is pertinent at this point to consider the definitions used to describe the amount of petroleum that remains in subterranean reservoirs.

Petroleum and heavy oil are resources; in particular, fossil fuel resources. A *resource* is the entire commodity that exists in the sediments and strata, whereas the reserves represent that fraction of a commodity that can be recovered economically. However, the use of the term reserves to describe the resource is subject to much speculation. In fact, it is subject to word variation. For example, reserves are classed as proved, unproved, probable, possible, and undiscovered.

Proven reserves are those reserves of petroleum that are actually found by drilling operations and are recoverable by means of current technology. They have a high degree of accuracy and are frequently updated as the recovery operation proceeds. They may be updated by means of reservoir characteristics, such as production data, pressure transient analysis, and reservoir modeling.

Probable reserves are those reserves of petroleum that are nearly certain but about which a slight doubt exists. *Possible reserves* are those reserves of petroleum with an even greater degree of uncertainty about recovery but about which there is some information. An additional term, *potential reserves,* is also used on occasion; these reserves are based upon geological information about the types of sediments where such resources are likely to occur, and they are considered to represent an educated guess. Then, there are the so-called *undiscovered reserves,* which are little more than figments of the imagination. The terms undiscovered reserves or *undiscovered resources* should be used with caution, especially when applied as a means of estimating reserves of petroleum. The data are very speculative and are regarded by many energy scientists as having little value other than unbridled optimism.

The term *inferred reserves* is also commonly used in addition to, or in place of, potential reserves. Inferred reserves are regarded as of a higher degree of accuracy than potential reserves, and the term is applied to those reserves that are estimated using an improved understanding of reservoir frameworks. The term also usually includes those reserves that can be recovered by further development of recovery technologies.

The differences between the data obtained from these various estimates can be considerable, but it must be remembered that any data about the reserves of petroleum (and, for that matter, about any other fuel or mineral resource) will always be open to questions about the degree of certainty. Thus, in reality (and in spite of the use of self-righteous word smithing), proven reserves may be a very small part of the total hypothetical and/or speculative amounts of a resource.

At some time in the future, certain resources may become reserves. Such a reclassification can arise as a result of improvements in recovery techniques, which may either make the resource accessible or bring about a lowering of the recovery costs and render winning of the resource an economical proposition. In addition, other uses may also be found for a commodity, and the increased demand may result in an increase in price. Alternatively, a large deposit may become exhausted and unable to produce any more of the resource, thus forcing production to focus on a resource that is lower grade but has a higher recovery cost.

More recently, the Society for Petroleum Engineers, with the American Association of Petroleum Geologists (AAPG), the World Petroleum Council (WPC), and the Society of Petroleum Evaluation Engineers (SPEE) (SPE, 2007), has developed a resource classification system (Figure 2–2) that moves away from systems in which all quantities of petroleum that are estimated to be initially-in-place are used. In these definitions, the quantities estimated to be initially-in-place are: (1) total petroleum-initially-in-place, (2) discovered petroleum-initially-in-place, and (3) undiscovered petroleum-initially-in-place. The recoverable portions of petroleum are defined separately as: (1) reserves, (2) contingent resources, and (3) prospective resources. In any case and whatever the definition, reserves are a subset of resources. They are those quantities of petroleum that are discovered (i.e., in known accumulations), recoverable, commercial, and remaining.

The *total petroleum-initially-in-place* is that quantity of petroleum (or heavy oil) that is estimated to exist originally in naturally occurring accumulations. The total petroleum-initially-in-place is, therefore, that quantity of petroleum that is estimated, on a given date, to be contained in known accumulations, plus those quantities already produced therefrom, plus those estimated quantities in accumulations yet to be discovered. The total petroleum-initially-in-place may be subdivided into *discovered petroleum-initially-in-place* and *undiscovered petroleum-initially-in-place*, with discovered petroleum-initially-in-place being limited to known accumulations.

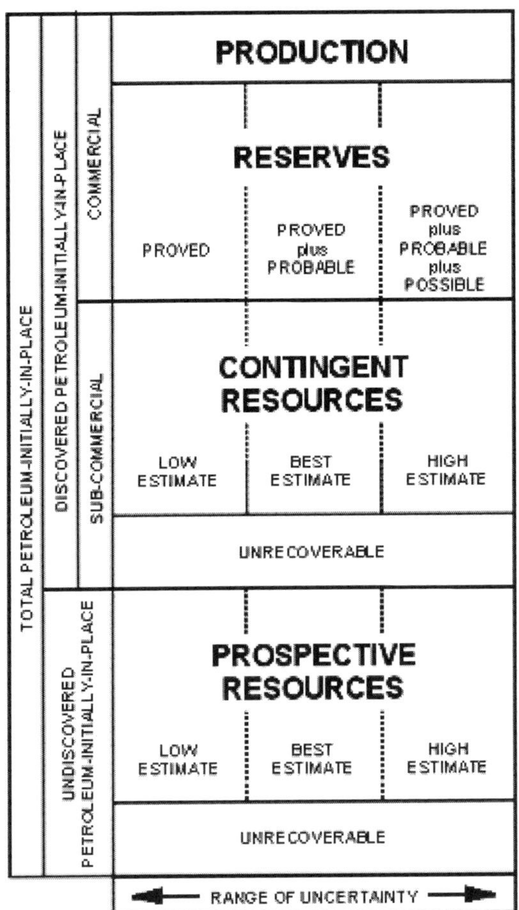

Figure 2–2 *Representation of resource estimation.*[a]

a. The horizontal axis represents the range of uncertainty in the estimated potentially recoverable volume for an accumulation, whereas the vertical axis represents the level of status/maturity of the accumulation. The vertical axis can be further sub-divided to classify accumulations on the basis of the commercial decisions required to move an accumulation towards production.

It is recognized that the quantity of petroleum-initially-in-place may constitute *potentially recoverable resources* since the estimation of the proportion that may be recoverable can be subject to significant uncertainty and will change with variations in commercial circumstances, technological developments, and data availability. A portion of those quantities classified as *unrecoverable* may become recoverable resources in the future as commercial circumstances change, technological developments occur, or additional data are acquired.

Discovered petroleum-initially-in-place is that quantity of petroleum that is estimated, on a given date, to be contained in known accumulations, plus those quantities already produced therefrom. Discovered petroleum-initially-in-place may be subdivided into *commercial* and *subcommercial* categories, with the estimated potentially recoverable portion being classified as *reserves* and *contingent resources,* respectively (as defined below).

Reserves are those quantities of petroleum that are anticipated to be commercially recovered from known accumulations from a given date forward. Estimated recoverable quantities from known accumulations that do not fulfill the requirement of commerciality should be classified as contingent resources (as defined below). The definition of commerciality for an accumulation will vary according to local conditions and circumstances and is left to the discretion of the country or company concerned. However, reserves must still be categorized according to specific criteria. Proven reserves will be limited to those quantities that are commercial under current economic conditions, whereas probable and possible reserves may be based on future economic conditions. In general, quantities should not be classified as reserves unless there is an expectation that the accumulation will be developed and placed on production within a reasonable timeframe.

Contingent resources are those quantities of petroleum that are estimated, on a given date, to be potentially recoverable from known accumulations but which are not currently considered as commercially recoverable. Some ambiguity may exist between the definitions of contingent resources and unproved reserves. This is a reflection of variations in current industry practice. If the degree of commitment is not such that the accumulation is expected to be developed and placed on production within a reasonable timeframe, however, the estimated recoverable volumes for the accumulation should be classified as contingent resources. Contingent resources may include, for example, accumulations for which there is currently no viable market, or accumulations for which commercial recovery is dependent on the development of new technology, or accumulations whose evaluation is still at an early stage.

Undiscovered petroleum-initially-in-place is that quantity of petroleum that is estimated, on a given date, to be contained in accumulations yet to be discovered. The estimated potentially recoverable portion of undiscovered petroleum-initially-in-place is classified as *prospective resources*.

Estimated ultimate recovery (EUR) is the quantity of petroleum that is estimated, on a given date, to be potentially recoverable from an accumulation, plus those quantities already produced therefrom. Estimated ultimate recovery is not a resource category, but it is a term that may be applied to an individual accumulation of any status/maturity (discovered or undiscovered).

Petroleum quantities classified as reserves, contingent resources or prospective resources should not be aggregated with each other without due consideration of the significant differences in the criteria associated with their classification. In particular, there may be a significant risk that accumulations containing contingent resources or prospective resources will not experience commercial production.

The *range of uncertainty* (Figure 2–2) reflects a reasonable range of estimated potentially recoverable volumes for an individual accumulation. Any estimation of resource quantities for an accumulation is subject to both technical and commercial uncertainties, and should, in general, be quoted as a range. In the case of reserves, and where appropriate, this range of uncertainty can be reflected in estimates for *proved reserves* (1P), *proved plus probable reserves* (2P), and *proved plus probable plus possible reserves* (3P) scenarios. For other resource categories, the terms *low estimate, best estimate,* and *high estimate* are recommended.

The term best estimate is used as a generic expression for the estimate considered to be the closest to the quantity that will actually be recovered from the accumulation between the date of the estimate and the time of abandonment. If probabilistic methods are used, this term would generally be a measure of central tendency of the uncertainty distribution. The terms low estimate and high estimate should provide a reasonable assessment of the range of uncertainty in the best estimate.

For undiscovered accumulations (*prospective resources*) the ranges will, in general, be substantially greater than the ranges for discovered accumulations. In all cases, however, the actual range will be dependent on the amount and quality of data (both technical and commercial) that is available for that accumulation. As more data becomes available for a specific accumulation (e.g., additional wells or reservoir performance data), the range of uncertainty in the estimated ultimate recovery for that accumulation should be reduced.

The low estimate, best estimate, and high estimate of potentially recoverable volumes should reflect some comparability with the reserves categories of proved reserves, proved plus probable reserves, and proved plus probable plus possible reserves, respectively. While there may be a significant risk that subcommercial or undiscovered accumulations will not achieve commercial production, it is useful to consider the range of potentially recoverable volumes.

In terms of actual numbers, on a worldwide basis, the produced conventional crude oil is estimated to be approximately 784 billion (784×10^9) bbl, with approximately 836 billion bbl remaining as reserves. It is also estimated that there are 180 billion bbl that remain to be discovered, with approximately 1 trillion (1×10^{12}) bbl yet to be produced. The annual depletion rate is estimated to be 2.6%.

2.3.2 The Real Numbers

One of the issues that arises when delineating and placing a value on heavy oil reserves is that because of the often unclear and confusing nomenclature, many estimates of heavy oil reserves also contain numbers related to the tar sand bitumen reserves. Until a unified system of nomenclature is accepted, there will always be differences between the various estimates, leading to publication of these differences in the relevant literature (Kovarik, 2003)

Venezuela has 47 to 76 billion barrels of proven reserves, according to oil industry and U.S. Department of Energy (DOE) estimates. The United States Geological Survey (USGS) puts Venezuelan reserves around the same level, at 48 identified and 110 ultimately recoverable. Venezuela claims 1.2 trillion (1.2×10^{12}) barrels of unconventional oil reserves in the supergiant heavy oil field stretching from the mouth of the Orinoco River near Trinidad down the east side of the Andes mountains. (Arcaya, 2001) The oil is located in a geosynclinal trough that is theorized to be continuous through the Falkland Islands, off the coast of Argentina. Only parts of the heavy oil field have been fully explored, but those parts have been estimated at some three to four trillion barrels of heavy oil in place, with perhaps one third recoverable using current technology.

Venezuela's oil reserves may almost triple in two years as oil companies develop the country's Faja, or heavy oil belt, according to Venezuelan President Hugo Chavez (Caribbean Net News 2008).

Proven reserves may rise to 171 billion barrels from the current 77 to 81 billion barrels by November 2007. When heavy oil and bitumen reserves are included in the total reserves, it is possible that Venezuela might supplant Saudi Arabia by 2008 as the country with the world's largest oil reserves. The BP Review of Energy Statistics (June, 2008) pegged Saudi Arabia's reserves at in excess of 250 billion barrels.

In addition, Venezuela's heavy oil deposits may rival Canada's tar sand deposits as an alternative source of oil to meet growing world fuel demand. Venezuela's heavy oil is high in sulfur, coke, and metals and needs more refining than conventional oil however. The estimated volume worldwide of technically recoverable heavy oil (434 billion barrels) and natural bitumen (651 billion barrels) in known accumulations is about equal to the remaining conventional (light) oil reserves (Table 2–2).

2.4 Production

Once oil is found in commercial quantities, there is a substantial delay until production begins, and such a delay can last several years in many offshore areas. After the initial discovery, further drilling and planning is necessary to determine the optimum way to exploit the find. In addition, the necessary infrastructure (pipelines, other supporting equipment, and including people) must be located and acquired. In many Arctic areas, the technology to allow environmentally acceptable recovery of oil is still being developed.

Production begins initially with the oil allowed to flow naturally from the high-pressure area underground to the surface. This natural recovery, called primary recovery, depends on reservoir pressure and the natural drive mechanism. The drive mechanism refers to the sources of energy within the reservoir that will assist production and properties of the oil, gas, and water found in relative proportions and locations.

Secondary recovery is another procedure used to boost recovery from a reservoir. This procedure consists of reinjection of either associated gas or water near the well to maintain underground pressure. In the past, secondary recovery techniques followed the use of primary

Table 2–2 Regional Distribution of Estimated Technically Recoverable Heavy Oil and Natural Bitumen in Billions of Barrels (bbl x 10^9)

Region	Heavy oil		Natural bitumen	
	Recovery factor	Technically recoverable bbl x 10^9	Recovery factor	Technically recoverable bbl x 10^9
North America	0.19	35.3	0.32	530.9
South America	0.13	265.7	0.09	0.1
W. Hemisphere	0.13	301.0	0.32	531.0
Africa	0.18	7.2	0.10	43.0
Europe	0.15	4.9	0.14	0.2
Middle East	0.12	78.2	0.10	0.0
Asia	0.14	29.6	0.16	42.8
Russia	0.13	13.4	0.13	33.7
E. Hemisphere	0.13	133.3	0.13	119.7
World		434.3		

recovery techniques; today these techniques may be used simultaneously to increase total production.

The recovery percentage of the oil in place recovered varies greatly from field to field. Thirty-three percent is sometimes used as an average figure, based upon experience in the U.S.; however, recovery in the East Texas oil field, which has a natural water drive, may exceed 75%. Production rates are limited by the drive mechanism and other geologic factors. In addition, economic and resource-maximization considerations (production at too fast a rate would lead to a reduction in total recovery) also influence production rates. A general planning guide for a country or region is that 1/15 of the recoverable oil remaining can be produced each year; this is referred to as the 15 to 1 reserves-to-production ratio. This planning guide is only a rough estimate and is, to some extent, a function of well density. Production from U.S., Canadian, and Australian fields, which are in a mature state of development, exceed the 1/15 ratio. In the United States, the figure is about 1/10.

With the increased price of oil since the 1973 embargo, other procedures have become economical for recovering additional oil from existing fields where primary and secondary techniques have already been used. This is especially true for the U.S. since many early U.S. fields were exploited ineffectively, causing a large portion of oil to be left in the ground. These additional techniques, called enhanced recovery techniques, consist of injecting heat or chemicals into the reservoir. Except for heat injection, these techniques are in the developmental stage. The success of these techniques varies greatly from field to field. A range of approximately 2% to 10% has been given as the additional percentage of oil in place that can be recovered in the U.S. using enhanced recovery techniques.

Since enhanced recovery techniques are expensive and require energy for their utilization, the extent of their use will depend upon the future price of oil and the price of the chemicals and steam required for their implementation. The production of heavy oil requiring steam for its primary recovery is not included in enhanced recovery estimates. However, heavy oil fields for which some production is possible without steam injection are included in enhanced recovery estimates. Thus, most heavy oil fields in California are included in enhanced recovery estimates.

2.5 Oil Pricing

Currently, oil is the primary energy source in the world (BP, 2007). For a century, the world has depended on low-cost oil to stimulate and maintain economic growth (Yergin, 1991). However, the ability to sustain the rate of economic growth is open to question. The volume of oil that can ultimately be recovered is subject to much speculation because of the uncertainties of reserve estimation, and this, in turn, affects the price of oil.

At this stage, it is appropriate to deal with the topic of *crude oil prices*. However, it is not the intent here to move into predictions of the future. Predictions of future events are always difficult—it is difficult to be correct. Everyone can justify with amazingly accurate 20/20 hindsight why his or her predictions were incorrect. These erstwhile mediums will use statistics to show that, after several rounds of mathematical manipulation, their predictions were very close to reality. even though the outcome bears no relationship to what really happened. It is easy to make statement that oil prices will continue to rise (after all, the pessimist is never disappointed). but the challenge is determining when and by how much.

Therefore, it is the aim of this section to forgo any predictions. It is, however, the purpose of this section to present a brief history of oil prices from which the reader can make his/her own predictions.

2.5.1 Oil Price History

Crude oil prices have seen wide price swings over the past decade, whether it is due to apparent shortage or oversupply. At the time of writing, prices are above $120 per barrel. Even when adjusted for inflation to current dollars, an average price per barrel bears little relationship to reality, especially when reality is a much higher price per barrel.

Historically (or, some might say, hysterically), crude oil prices varied from $2.50 to $3.00 from 1948 through the end of the 1960s. The price rose from $2.50 in 1948 to about $3.00 in 1957. From 1958 to 1970, prices were stable at about $3.00 per barrel. The Organization of Petroleum Exporting Countries (OPEC) was formed in 1960 with five founding members: Iran, Iraq, Kuwait, Saudi Arabia, and Venezuela. By the end of 1971, six other nations (Qatar, Indonesia, Libya, United Arab Emirates, Algeria, and Nigeria) had swelled the membership ranks of OPEC.

Throughout this period, the petroleum-exporting countries found increasing demand for their crude oil. In 1972, the price of crude oil was about $3.00 per barrel, and by the end of 1974, the price of oil had quadrupled to over $12.00. The Yom Kippur War started with an attack on Israel by Syria and Egypt on October 5, 1973. Many countries in the western world showed strong support for Israel, and as a result, several of the Middle Eastern oil-exporting nations imposed an embargo on those countries by decreasing oil exported.

From 1974 to 1978, the price of crude oil was relatively flat, ranging from $12.21 per barrel to $13.55 per barrel. When adjusted for inflation, the price over that period of time exhibited a moderate decline. Then, events in Iran and Iraq (the overthrow of the Shah of Iran and the Iran-Iraq War) led to another round of crude oil price increases. Crude oil prices rose to $35 per barrel in 1981.

The higher prices (along with other factors, usually involving energy conservation) resulted in increased exploration and production in the non-OPEC world. In mid-1985, oil prices were linked to the spot market for crude oil, and by early 1986 (with increased production by some OPEC members), crude oil prices moved downward to $8 to $10 per barrel. The price of crude oil rose again in 1990 with the Iraqi invasion of Kuwait and the ensuing Gulf War, but following the war, crude oil prices entered a steady decline. The price cycle then turned up, and from 1990 to 1997, world oil consumption increased by six million barrels per day. The price increases came to a rapid end when, due to a downward trend in several Asian economies, higher OPEC production sent prices downward.

A low point was reached in January 1999, after increased oil production from Iraq coincided with the Asian financial crisis, which reduced demand. Oil prices then rapidly increased, more than doubling by September 2000. Prices then fell until the end of 2001, before steadily increasing, reaching $40 to $50 per barrel by September 2004. In October 2004, the price of crude oil exceeded $53 per barrel, and for December delivery, it exceeded $55 per barrel. Crude oil prices surged to a record high, above $60 a barrel in June 2005, sustaining a rally built on strong demand for gasoline and diesel and on concerns about refiners' ability to keep up. This trend continued and crude oil prices surged to $147 per barrel before decreasing to less than $50 per barrel. This surge does not bode well for the future of crude oil prices, but opens the door to the increased use of cheaper heavy oil.

2.5.2 Pricing Strategies

Crude oil is of little use before refining and is traded for the final petroleum products that consumers demand. The intrinsic properties of crude oil determine the mix of final petroleum products. The two most important qualities of crude oil are viscosity (thickness or density) and sulfur content.

Light crude oils contrast with heavy crude oils, which have a low share of light hydrocarbons and require a much more complex refining process than distillation (such as coking and cracking) to produce similar proportions of the more valuable petroleum products. Sulfur is an undesirable property of crude oil, and refiners make heavy investments in order to remove it.

Since the type of crude oil has a bearing on refining yields, different types of crude oil fetch different prices. Crude oils that yield a higher proportion of the more valuable final petroleum products and require simple refining processes (the light/sweet crude variety) usually command a premium over those that yield a lower percentage of the more valuable petroleum products and require more complex refining processes (the heavy/sour crude variety). For example, West Texas intermediate crude oil, which is a sweet/light crude oil, trades at a premium compared to any crude oil that is considered a sour/heavy crude variety. The price differential can reach very high levels, and there has been a rise in the price differential during recent years.

However, differences in quality are not the only determinants of oil price differentials, and hence differentials are not constant over time. Crude oil price differentials are influenced by a wide array of factors and are highly volatile. For example, changes in the prices of different petroleum products (or the gross product worth) lead to changes in crude oil differentials. Another factor that may influence the differentials is the heating season and whether the season turns out to be colder than expected, which suggests that movements in crude oil price differentials are likely to exhibit a seasonal behavior. Factors outside the oil sector can also have an influence.

References to the oil price are usually either references to the spot price of either West Texas (light) crude oil traded on the New York Mercantile Exchange (NYMEX) for delivery in Cushing, Oklahoma or the price of Brent crude oil traded on the International Petroleum Exchange (IPE) for delivery at Sullom Voe, in the Shetland Islands, Scotland.

The price of a barrel of oil is highly dependent on both its grade (which is determined by factors such as its specific gravity or API and its sulfur content) and location. The vast majority of oil will not be traded on an exchange but on an over-the-counter basis, typically with reference to a marker crude oil grade that is typically quoted via the pricing agency IPE claim that 65% of traded oil is priced off their Brent benchmarks. The Energy Information Administration (EIA) uses the Imported Refiner Acquisition Cost, the weighted average cost of all oil imported into the United States as the *world oil price*.

Heavy crude oil provides an interesting situation for the economics of petroleum development. The resources of heavy oil in the world are more than twice the resources of conventional crude oil. On one hand, heavy crude oils are often priced at a discount to lighter ones due to increased refining costs and high sulfur content, and the increased viscosity and density also makes production more difficult. On the other hand, large quantities of heavy crude oil exist in the Americas, such as in Canada (particularly in Alberta and Saskatchewan), the United States (particularly Northern California), and Venezuela (particularly in the Orinoco). The relatively shallow depth of these reservoirs (often less than 3000 feet) contributes to lower drilling costs, but recovery costs (steam generation) are relatively high.

Large discounts increased the attractiveness of heavy crude oils to refineries, which responded by increasing their imports of heavy crude oil and increasing the production of refined petroleum products to meet the rise in demand. Thus, the incremental demand growth for light petroleum products, mainly gasoline, can be met by imports of cheaper heavier crude oils. However, the use of more heavy crude oil instead of conventional light crude oil may be detrimental to the ability of many refineries to cope with the increased amount of high-sulfur feedstocks. It may also be a burden on the refining capacity when the specter of more-feedstock-for-the-same-amount-of-gasoline raises its head.

2.5.3 The Role of Heavy Oil in the Future

The Hubbert peak theory assumes that oil reserves will not be replenished (i.e., that abiogenic replenishment is negligible) and predicts that future world oil production must inevitably reach a peak and then decline as these reserves are exhausted. Controversy surrounds the theory since predictions for the time of the global peak are dependent on the past production and discovery data used in the calculation.

For the United States, the prediction turned out to be correct. After the U.S. oil production peaked in 1971 and thus lost its excess production capacity, OPEC was able to manipulate oil prices. Since then, oil production in several other countries has also peaked. However, for a variety of reasons, it is difficult to predict the oil peak in any given region. Based on available production data, proponents have previously (and incorrectly) predicted the peak for the world to be in years 1989, 1995, or in the 1995 to 2000 period. Other predictions have chosen 2007 and beyond for the peak of oil production.

In summary, the petroleum industry is indeed at the verge of a major decision period with the onset of processing of high volumes of heavy crude oil and residua. Several technology breakthroughs have made this possible, but many technical challenges still remain.

More important, several trends that should have been established in the wake of decreasing crude oil prices have never been put into practice. Some would argue that the periods of oil price decline were the impetus to development of better technology and expertise. They would also argue that politicians at various levels of government have failed to recognize the need for a measure of energy independence through the development of alternate resources as well through the development of technologies that would assist in maximizing domestic oil recovery.

2.6 References

Arcaya, I. 2001. Venezuela and the United States: A Four-Pillar Strategy for Energy Security. Remarks to the Business Council for International Understanding Petroleum Club of Houston, Houston, Texas. July 24.

Gold, T. 1984. Scientific American. 251(5):6.

Gold, T. 1985. Annual Reviews of Energy. 10:53.

Gold, T., and Soter, S. 1980. Scientific American. 242(6):154.

Gold. T., and Soter, S. 1982. Energy Exploration Exploitation. 1(1):89.

Gold. T., and Soter, S. 1986. Chemical Engineering News. 64(16):1.

Head, I.M., Jones, D.M., and Larter, S.R. 2003, Biological Activity in the Deep Subsurface and the Origin of Heavy Oil. Nature. 426(20):344–352.

International Energy Agency, 2005. Resources to Reserves: Oil and Gas Technologies for the Energy Markets of the Future. International Energy Agency, Paris, France.

Kenney, J.F., Shnyukov, A.Y.F., Krayushkin, V.A., Karpov, I.K., Kutcherov, V.G., and Plotnikova, I.N. 2001. Dismissal of the Claims of a Biological Connection for Natural Petroleum. Energia. 22(3):26–34

Kovarik, W. 2003. The Oil Reserve Fallacy: Proven Reserves Are Not A Measure Of Future Supply. http://www.runet.edu/~wkovarik/oil/3unconventional.html.

Larter, S, Whilhelms, A., Head, I., Koopmans, M., Aplin, A., Di Primio, R., Zwach, C., Erdmann, M., and Telnaes, N. 2003. The Controls on the Composition of Biodegraded Oils in the Deep Subsurface—Part 1: Biodegradation Rates in Petroleum Reservoirs. Organic Geochemistry. 34(3):601–613.

Larter, S., Huang, H., Adams, J., Bennett, B., Jokanola, O., Oldenburg, T., Jones, M., Head, I., Riediger, C. and Fowler, M. 2006. The Controls in the Composition of Biodegraded Oils in the Deep Subsurface—Part II: Geological Controls on Subsurface Biodegradation Fluxes and Constraints on Reservoir-Fluid Property Prediction. American Association of Petroleum Geologists Bulletin, 90(6):921–938.

Meyer, R.F., and Attanasi, E.D. 2003. Heavy Oil and Natural Bitumen—Strategic Petroleum Resources. Fact Sheet 70-03. U.S. Geological Survey, Washington, DC. http://pubs.usgs.gov/fs/fs070-03/fs070-03.html.

Meyer, R.F., and Attanasi, E.D. 2004. Natural Bitumen and Extra Heavy Oil. World Energy Council 2004 Survey of Energy Resources. Elsevier, Amsterdam, The Netherlands.

Meyer, R.F., Attanasi, E.D., and Freeman, P.A. 2007. Heavy Oil and Natural Bitumen Resources in Geological Basins of the World: Open-File Report 2007-1084. U.S. Geological Survey, Washington, DC. http://pubs.usgs.gov/of/2007/1084/.

Morehouse, D.F. 1997. The Intricate Puzzle of Oil and Gas Reserves Growth. Natural Gas Monthly. U.S. Energy Information Administration, Washington, DC.

Osborne, D. 1986. Atlantic Monthly. February. p. 39.

Society for Petroleum Engineers, 2007. Petroleum Resources Management System. Society for Petroleum Engineers, Richardson, Texas.

Speight. J.G. 2007. The Chemistry and Technology of Petroleum. 4th Edition. CRC-Taylor and Francis Group, Boca Raton, Florida.

Szatmari, P. 1989. Bull. Am. Assoc. Petrol. Geol. 73(8):989.

Yergin, D. 1991. The Prize: The Epic Quest for Oil, Money, and Power. Simon & Schuster, New York.

Caribbean Net News. August 11, 2006. Venezuala Oil Reserves to Soar, Boosted by Heavy Oil. http://www.caribbeannetnews.com/cgi-script/csArticles/articles/000027/002726.htm

CHAPTER 3

RESERVOIRS AND RESERVOIR FLUIDS

The reservoir characteristics are usually the focal point of recovery operations. Detailed reservoir characterizations are the foundation for future decisions, such as selecting the right time and follow-up production method to enhance overall recovery. The improved ability to predict how a reservoir will behave benefits the project.

However, reservoir fluids (i.e., the fluids—including gases and solids—that exist in a reservoir) are of equal importance. The fluid type must be determined very early in the life of a reservoir (often before sampling or initial production) because fluid type is the critical factor in many of the decisions that must be made about producing the fluid from the reservoir. Reservoir-fluid properties play a key role in the design and optimization of injection/production strategies and surface facilities for efficient reservoir management. Inaccurate fluid characterization often leads to high uncertainties in in-place-volume estimates and recovery predictions, and hence it affects asset value.

Indeed, in addition to reservoir evaluation, evaluating reservoir fluids is an important aspect of prospect evaluation, development planning, and reservoir management. Although many different types of fluid exist, the composition of the reservoir fluids provides important information that can influence the recovery process. Prior to production, these measurements will represent static data, but once production commences, dynamic data will become available.

The reservoir fluid data is an important compliment to the pressure data and has the advantage of providing a direct measure of reservoir fluid continuity. Many different measurements of the reservoir fluid are available, both at reservoir and atmospheric conditions, but it has become increasingly common to apply some type of detailed oil fingerprinting for this purpose. Typically, the analysis of reservoir pressures and the oil fingerprinting data will give similar results. In some situations, however, this is not the case, and it can be a source of confusion in determining the true nature of the reservoir architecture. When these situations occur, it is often because the processes that control the reservoir pressures and the timing of their development are different from those controlling the reservoir fluid filling history. In this paper we will show several case studies illustrating the complimentary nature of reservoir pressure and fluid composition data. We will also discuss geologic settings where these two types of data give divergent views of reservoir continuity and offer an explanation for this behavior.

Reservoir fluids, including heavy oil, vary greatly in composition. In some fields, the fluid is in the gaseous state, and in others, it is in the liquid state. Gas and liquid frequently coexist in a reservoir. The rocks which contain these reservoir fluids also vary considerably in composition and in physical and flow properties, and this can serve to complicate the sampling procedure. Other factors, such as producing area, height of the column of hydrocarbon fluid, fracturing or faulting, and water production also serve to distinguish one reservoir from another. The combination of all these factors affects the choice of sampling methods and preparations for sampling.

The mobility of reservoir fluids and the geometry or structure of the reservoir influences recovery rates. In order to correctly specify the necessary downhole equipment, it is important to understand those fluid properties.

3.1 Reservoirs

A *reservoir* is a subsurface, porous, permeable rock body or formation that has the capability to store and transmit fluids: oil, water, and gas. A reservoir is characteristically large and extensive in volume and capacity and is created by the sequential steps of deposition, conversion, migration, and entrapment.

Deposition occurs when organic-rich sediment occurring in areas of high primary productivity is deposited and settles due to low energy currents. Such environments include rivers deltas, estuaries, and continental shelves. Further deposition of sediment results in build-up of layers of sediment. For example, silts or clays atop previously organic-rich sediment prevent oxygen from getting into the system, effectively preserving organic matter. Further burial of the organic matter occurs due to more deposition, and pressures and temperatures of the layers increase due to increasing depth and the geothermal gradient 25° to 30°C/km or 15°F/1000 ft. This results in the conversion of organic matter to petroleum.

Conversion is a breakdown of the organic matter that involves three main stages—diagenesis, catagenesis, and metagenesis—to form kerogen. Kerogen is the high-molecular-weight organic residue found in many types of sediment, particularly in oil shale.

Migration relates to the movement of oil or gas from its source rock, where it was formed, to a reservoir rock. Porosity and permeability of the rocks control the migration of the hydrocarbon.

Entrapment occurs when the hydrocarbon migrates into a formation that is effectively sealed due to a change in the lithology of the formation. For example, shale is commonly porous, but because of their fine grain size, they have very high capillary forces that prevent fluid flow.

Oil (conventional or heavy) cannot be retained as an accumulation unless there is a trap. This requires that the boundary between the cap rock or other sealing agent and the reservoir rock generally be convex upward, but the exact form of the boundary varies widely. The simplest forms are the flat-lying convex lens, the anticline, and the dome, each of which has a convex upper surface (Figure 3–1). Many oil and gas accumulations are trapped in anticlines or domes, structures that are generally more easily detected than some other types of traps.

The most common reservoir rocks are sandstone, limestone, and dolomite. The five basic elements of a reservoir system include: (1) source rock—the rock containing the organic material that converted into petroleum reservoir fluid, (2) a migratory pathway, (3) reservoir rock—the rock that can store and yield the fluid (that has sufficient porosity and permeability), (4) a seal—the impermeable cap rock that prevents the upward escape of petroleum to the earth's surface, and (5) a trap—the physical arrangement or space that prevents migration of reservoir fluid, whereby source, reservoir rock, basement rock, and seal are arranged to trap petroleum.

The reservoir rock is a porous medium made up of pores conventionally envisaged as irregularly shaped holes some 1 to 100 μm in length and diameter that are connected to maybe six other pores. There can be some 10^6 pores in one cubic centimeter of rock and possibly over 10^{22} pores in a typical reservoir.

3.1.1 Structural Traps

Reservoirs (i.e., oil traps) are created by structural deformation of the geological strata; these traps are formed by tectonic processes after the deposition of the beds involved. There are three basic forms of a *structural trap* in petroleum geology: (1) an anticline trap, (2) a fault trap, and (3) a salt dome trap (Figures 3–1, 3–2, and 3–3). The anticline is a typical structural trap (Figure 3–1) that is produced by compressional folding, by uplift, and by drape over older tectonically created features. An anticline is an example of rocks which were previously flat but have been bent into an arch. The rocks have been folded or bucked into the form of a dome, and hydrocarbons accumulate in the hinge area of an anticline.

Another form of trap is the *stratigraphic trap,* which is formed when other beds seal a reservoir bed or when the permeability changes within the reservoir bed itself (through a change in facies, i.e., a change due to the presence of rock with characteristics different to those of the reservoir rock or a change in lithology). The distinction between a *structural trap* and a *stratigraphic trap* is often blurred. For example, an anticlinal trap may be related to an underlying buried limestone reef. Beds of sandstone may wedge out against an anticline because of depositional variations or intermittent erosion intervals. Salt domes, formed by flow of salt at substantial depths, also have created numerous traps that are both a structural trap and a stratigraphic trap.

In order to become a reservoir, the rock must possess *fluid-holding capacity* (*porosity*) and also *fluid-transmitting capacity* (*permeability*). A variety of different types of openings in rocks are responsible for these properties in reservoir rocks. The most common are pores between the grains of the reservoir rock are made, the cavities inside fossils, and openings formed by solution or fractures. The relative proportion of the different kinds of openings varies with the rock type, but pores usually account for the bulk of the storage space. The *effective porosity* for oil storage results from continuously connected openings. These also provide the permeability. Although a rock must be porous to be

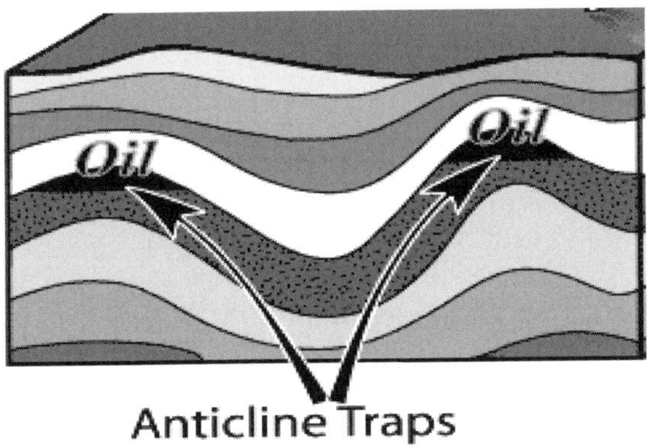

Figure 3–1 *Anticlinal traps.*

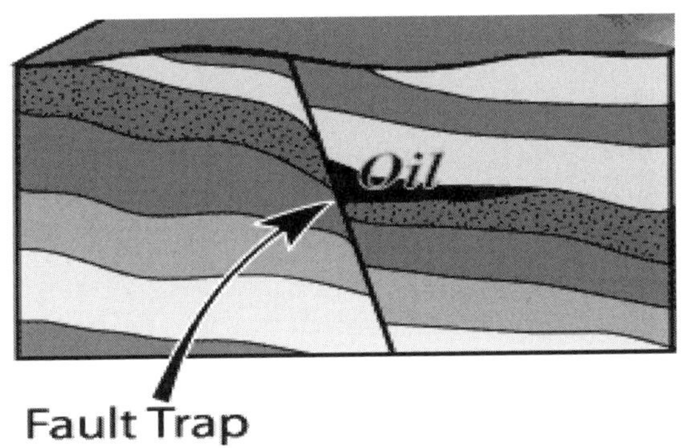

Figure 3–2 *A fault trap.*

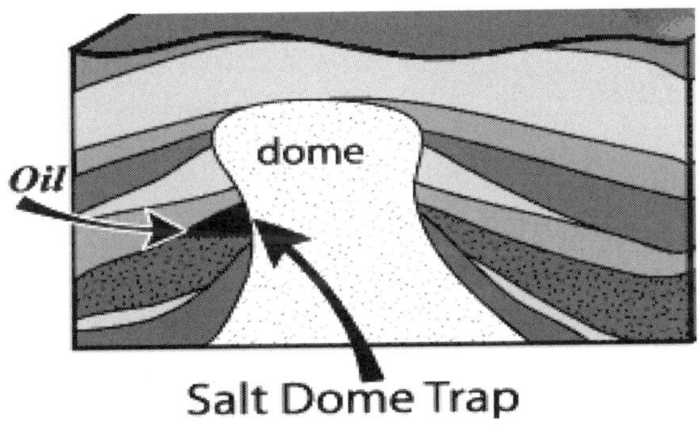

Figure 3–3 *A salt dome trap.*

permeable, there is no simple quantitative relationship between the porosity and permeability.

Reservoir rocks tend to show far greater variations in permeability than in porosity. In addition, the values of these two properties, as measured on core samples from reservoir rocks, are not always identical with the values indicated for the rock in bulk underground. The differences arise from the unrepresentative nature of cores, especially when there are wide variations in the sizes of the openings in the rocks and irregularities in their distribution. Porosity is generally in the range of 5% to 30%; whereas permeability is commonly between 0.005 darcy (5 millidarcies) and several darcies (several thousand millidarcies), as measured on small samples. It should be noted that pores may be, at best, only a millimeter or so in width, whereas fossil and solution cavities may sometimes be 30 to 50 times wider. Many joints and fractures are probably only a millimeter across, although they may extend for considerable distances.

For the purposes of defining a reservoir and understanding reservoir behavior as it relates to heavy oil recovery, the process involves using data from a variety of pore-core-reservoir data and laboratory-field sources to improve understanding reservoirs (Kovscek, 2002).

3.1.2 Heterogeneity

In addition to the understanding of the petrophysics of the reservoir, oil recovery requires an understanding of displacement and flow through porous media (Dawe, 2004). However, flow through porous media is complicated. Inside a reservoir there can be displacements and miscible and/or immiscible flow, with one, two, or sometimes three mobile phases (oil, gas, and water) (Grattoni and Dawe, 2003). Understanding the physics of displacement is important for the correct interpretation of laboratory core data, for the assessment of quantities and position of residual oil, and for reservoir simulation.

Heterogeneity in the form of layers, lenses, cross-beds, and quadrants can have a profound effect on fluid displacement patterns. Even modest changes in rock permeability give rise to distortions in displacement profiles and disperse the streamlines and lower sweep efficiencies and recovery.

Physically, hydrocarbon reservoirs themselves are complicated geological heterogeneous bodies. They are not the homogeneous porous media that are often envisaged on paper and used in calculations. Heterogeneity means that a specific property of interest varies vertically and longitudinally within the reservoir (Dawe, 2004), much like the coal in a seam that varies in composition from one part of the seam to another (Speight, 1994). For example, well-log and core-analysis reports show that all reservoirs are heterogeneous with rock properties (porosity, saturation, etc.) varying within the reservoir. As we will show, permeability heterogeneities cause variations in the fluid movements compared to the equivalent homogeneous system. Often, the effects of heterogeneities are not well accounted for at the planning stage of an operation, and they only become evident when it may be too late and water has started to be produced before the predicted time (Dawe, 2004). An understanding of the movement of fluids within heterogeneous porous media is therefore fundamental to petroleum production and its efficient management.

The heterogeneities in reservoirs that need to be studied are those that interfere with the flow of fluids. Geological variations that do not create flow-pattern changes are not important for flow. The heterogeneities can be permeability or wettability variations (Caruana and Dawe, 1996a, 1996b; Dawe, 2004). The effects of layer thickness, permeability contrast, angle of layer to flow direction, mobility ratio, wettability, and flood rate have been examined. Each of these parameters influences the displacement profiles and disperses the flood front. It is more than likely that wettability of the reservoir rock by heavy oil (particularly adsorption of the polar constituents) can have major effects on heavy oil recovery.

Wettability of reservoirs rocks is the tendency of one fluid to spread on or adhere to a solid surface in the presence of other immiscible fluids. It is determined by complex *interface boundary conditions* acting within the pore space of sedimentary rocks. In general, at least one of the two immiscible fluids in a porous medium will be the wetting phase. When the system is in equilibrium, the wetting fluid will completely occupy the smallest pores and be in contact with the majority of the rock surface. The nonwetting fluid will occupy the center of the large pores and form globules that extend over several pores. When the rock is water wet, there is a tendency for water to occupy the small pores and to contact the majority of the rock surface. When the rock, which is preferentially in contact with oil, is *oil wet,* the oil will occupy the small pores and contact the majority of the rock surface. The minerals present in reservoir rocks are generally known as being intrinsically *hydrophilic,* i.e., preferentially water wet.

Wettability is generally considered to be one of the most important parameters influencing saturation, distribution, and flow of fluids in porous media. Knowledge of the wettability of reservoir rock is important to petroleum engineers and geologists. For example, a waterflood on a strongly oil-wet rock is much less efficient than one in water-wet rock (Anderson, 1986).

3.2 Classes of Fluids

In order to produce a wellhead product from a reservoir, there has to be flow of the fluids to the wellbores through the heterogeneous porous media of the reservoir. Fluid movements within the reservoir are governed by the local fluid potential gradients and reservoir effective permeability, the injection and production points, and the fluid viscosities. Fluid flow is also governed by the character of the fluids in the reservoir.

In the current context, reservoirs usually contain three main fluids: (1) natural gas, (2) oil, and (3) water, with minor constituents being acid gases (carbon dioxide and hydrogen sulfide). These components will vary greatly in combination and proportion within each reservoir. In the case of a heavy oil reservoir, the amount of gas will be substantially less than would be found in a conventional oil reservoir.

Reservoir fluids vary greatly in composition and chemical properties. As mentioned already, fluids exist in solid, liquid, and gaseous states, and gas and liquids often coexist in a given reservoir. The distribution of the fluids in a reservoir rock is dependent on the densities of the fluids as well as on the properties of the rock. If the pores are of uniform size and evenly distributed, there are: (1) an upper zone, where the pores are filled mainly by gas (the gas cap); (2) a middle zone, where the pores are occupied principally by oil with gas in solution; and (3) a lower zone, where the pores are filled by water. A certain amount of water (approximately 10% to 30%) occurs along with the oil in the middle zone. There is a transition zone from the pores occupied entirely by water to the pores occupied mainly by oil in the reservoir rock. The thickness of this zone depends on the densities and interfacial tension of the oil and water as well as on the sizes of the pores. Similarly, there is some water in the pores in the upper gas zone, which has at its base a transition zone from pores occupied largely by gas to pores occupied mainly by oil.

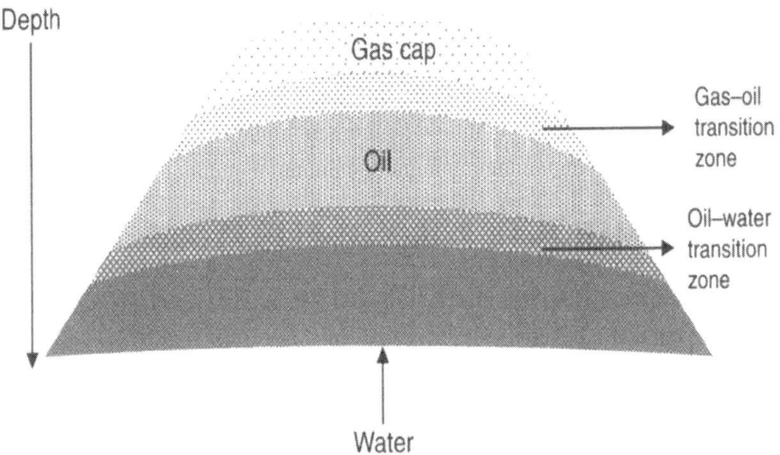

Figure 3–4 *Representation of the zones in a reservoir.*

The water found in the oil and gas zones (Figure 3–4) is known generally as *interstitial water.* It usually occurs as collars around grain contacts, as a filling of pores with unusually small throats connecting with adjacent pores, or (to a much smaller extent) as wetting films on the surface of the mineral grains when the rock is preferentially wet by water. The water may occur as wetting films or collars around the sand grains as well as in some completely filled pores. The three-dimensional network allows continuity to exist for the hydrocarbons by means of connections on every side of the sand grains. The so-called gas-oil and oil-water contacts are generally horizontal but have been known to exist as a very gentle incline. On occasion, part of an accumulation of the oil or gas has its lower boundary marked not by the water-bearing zone of the reservoir rock but by an adjacent sealing rock that has characteristics similar to those of the cap rock. When the pressure and temperature conditions are suitable in relation to the proportions and the nature of the gas and oil, there may be no gas cap but only oil, with dissolved gas overlying the water.

Heavy oil has an API gravity between 10° and 20° (Chapter 1), is more viscous than conventional petroleum, and has the commonality of being (or having the ability to be) in the liquid state and, therefore, having mobility in the reservoir. It can be recovered from a reservoir by the use of conventional (including enhanced) oil recovery techniques.

The water (brine) produced with oil has been trapped with the oil and is brought to the surface along with oil. Because the water has been in contact with the oil, it contains some of the chemical characteristics of the formation and the oil itself. Oil and gas wells produce more water than oil (7 bbl/1 bbl oil in some fields). The composition (salt content) of coproduced water determines the need for antiscaling additives. There are strict regulations to limit disposal and beneficial use options as well as environmental impacts that pertain to oil field waters.

Reservoirs contain complex fluid mixtures whose behavior is strongly dependent on chemical make-up. Heavy oil is a fluid that is also a multicomponent mixture, composed of non-hydrocarbons and a variety of hydrocarbons, especially of the alkane series. Typical hydrocarbons encountered in heavy oil are the higher-boiling hydrocarbons. The amount depends on the original source materials and the maturation pathways; volatile hydrocarbons boiling lower than C_{12} are not present in a ready abundance.

Reservoir temperatures may vary up to 90°C or more, while surface conditions are around 20°C. Pressure can vary from its atmospheric value (or lower in the case of vacuum distillation) to a number in the hundred million pascals (Pa). Within such an ample range of conditions, hydrocarbon fluids undergo severe transformations and exist as a single phase (gas, liquid, or solid) or coexist in several forms (liquid plus gas, solid plus liquid, vapor plus solid, or even in liquid-plus-liquid combinations). Understanding how hydrocarbon fluids interact with and react to their thermodynamic surroundings is essential to adequately analyze systems of interest. One of the most useful phase-behavior visualizations is the pressure-temperature (p-T) diagram or p-T envelope. Each envelope represents a thermodynamic boundary separating the two-phase conditions (inside the envelope) from the single-phase region (outside).

The correct identification of the type of hydrocarbon fluid is critical for the proper design and development of the correct production strategy for the field under consideration. This identification is critical for proper hydrocarbon reservoir modeling as well. When a fluid is assumed to behave as heavy oil, it is assumed that its behavior is complex, and selecting an appropriate model for behavior becomes a limiting factor.

Reservoir fluids are brought to the surface as a mixture of oil, gas, and water, which is sent to a surface production facility before it can be disposed or sold to an industrial customer (e.g., a refinery). A surface production facility is the system in charge of the separation of the well stream fluids into their three single-phase components—oil, gas, and water—and of their transport and processing into marketable products and/or their disposal in an environmentally acceptable manner. Once separated, the oil, natural gas, and water follow different paths. Water is typically reinjected for reservoir pressure maintenance operations. The oil usually goes through a process of dehydration, which removes basic sediments. Hydrocarbon fluids are assumed to comprise two components: stock tank oil and surface gas.

3.3 Evaluation of Reservoir Fluids

The successful exploitation of a heavy oil reservoir depends as much on understanding the fluid properties of the reservoir as it does on knowing the geology of the reservoir itself. Chemical differences between heavy oil and conventional oil ultimately affect their properties, such as viscosity. In fact, heavy oil properties influence every aspect of development of a heavy oil prospect. Technology that was developed for conventional oil does not usually address the issues of producing heavy oil. In fact, every new heavy oil development project eventually requires some form of enhanced oil recovery, which generally means steam, solvents, or a combination of both. Without enhanced oil recovery, the recovery factor from what the industry calls *cold production* (Chapter 6) might be as little as 1% and often no more than 10%. With thermal recovery, typical rates run from 30% to 70%. In thermal processes, however, the cost of generating the steam is typically the single greatest operating expense. Other techniques, such as injecting slugs of water alternating with gas (WAG), are less efficient than thermal recovery, but also less expensive.

Thus, reservoir-fluid pressure-volume-temperature (PVT) properties are critical for efficient reservoir management throughout the life of the reservoir, from discovery to abandonment (Honarpour et al., 2006; Nagarajan et al., 2007). In fact, reliable data related to the properties of in situ fluids are essential for the determination of in-place volumes and recovery-factor calculations as well as for technical evaluation of reservoir-development-depletion plans. Fluid characterization and distribution within the reservoir help in defining reservoir continuity and communication among various zones.

Reservoir-fluid characterization consists of several key steps: (1) acquisition of representative samples, (2) identification of reliable service laboratories to perform PVT measurements, (3) implementation of quality assurance/quality control (QA/QC) procedures to ensure data quality, and (4) development of mathematical models to capture fluid-property changes accurately as functions of pressure, temperature, and composition. The fluid type and production processes dictate the type and the volume of required fluid data.

3.3.1 Sampling Methods

The main objective of a successful sampling campaign is to obtain representative fluid samples for determining properties of the fluid (Speight, 2001, 2002). Adequate volumes should be collected for plant and process analysis, geochemical analysis for fluid-source identification and reservoir continuity, and crude assay for refinery processes (API, 2003). The sampling program should focus on selecting an appropriate sampling method and developing sound sampling, sample-transfer, and QC procedures. In addition, sample character and specific sampling issues should be addressed in the form of a sample history that details the acquisition, storage, and text carried out on the sample (Speight, 2001, 2002).

Proper management of production from a reservoir can maximize the recovery of the hydrocarbon fluids (gas and oil) originally in the reservoir. Developing proper management strategies requires accurate knowledge of the characteristics of the reservoir fluid. Practices are recommended herein for obtaining samples of the reservoir fluid, from which the pertinent properties can be determined by subsequent laboratory tests.

The objective of reservoir fluid sampling is to collect a sample that is representative of the fluid present in the reservoir at the time of sampling. If the sampling procedure is incorrect or if samples are collected from an improperly "conditioned" well, the resulting samples may not be representative of the reservoir fluid. An unrepresentative sample may not exhibit the same properties as the reservoir fluid. The use of fluid property data obtained from unrepresentative samples, however accurate the laboratory test methods, may result in errors in reservoir management. Poor planning can also result in incomplete data being taken during the sampling program. Incomplete data can make it difficult or impossible for laboratory personnel to perform and interpret tests that provide accurate and meaningful fluid property information.

As mentioned, reservoir fluids found in gas and oil fields around the world vary greatly in composition. The rocks which contain these reservoir fluids also vary considerably in composition as well as in physical and flow properties. In certain cases, this can serve to complicate the sampling procedure. Other factors, such as producing area, height of the column of hydrocarbon fluid, fracturing or faulting, and water production also serve to distinguish one reservoir from another. The combination of all these factors affects the choice of sampling methods and preparations for sampling.

When a reservoir is relatively small, a properly taken sample from a single well can be representative of the fluid throughout the entire reservoir. For reservoirs which are large or complex, samples from several wells and/or depths may be required. Significant variations in fluid composition often occur in very thick formations, in really large reservoirs, or in reservoirs subjected to recent tectonic disturbances. Additional sampling during the later life of a reservoir is not uncommon because production experience can show that the reservoir is more complex than earlier information indicated.

Methods for sampling reservoir fluids fall into two general categories: subsurface sampling or surface sampling. As the names imply, each category reflects the location at which the sampling process occurs. Subsurface sampling may also be referred to as downhole or bottomhole sampling.

The two commonly used sampling methods are bottomhole sampling and surface sampling. Bottomhole sampling attempts to capture samples close to reservoir conditions, whereas surface sampling aims at capturing gas and oil samples from the separator under stable flow conditions. Separator fluids then are recombined at a measured producing gas/oil ratio (GOR) to prepare representative reservoir fluid.

In bottomhole-sampling operations, adequate cleaning of near-wellbore regions and controlled drawdown are critical for obtaining uncontaminated representative samples (Witt and Crombie 1999). Controlled drawdown helps avoid two-phase flow in the reservoir. Downhole fluid analyzers are used to monitor sample contamination and ensure single-phase flow prior to sample capture. In surface sampling operations, proper well conditioning with minimum drawdown is the key to acquiring high-quality samples. Well conditioning requires that the well be flowed at an optimum rate for an extended period of time with a stable producing GOR, but it must be recognized that sample quality and separator efficiency introduce uncertainties in the quality of the fluids.

Modern open-hole wireline formation testers now provide the means of recovering representative samples before the effects of subsequent production take place. Selection of one particular method over another is influenced by the type of reservoir fluid, the producing characteristics and mechanical condition of the well, the design and mechanical condition of the surface producing equipment, the relative expense of the various methods, and safety considerations. Detailed descriptions of recommended sampling methods are presented in subsequent sections of this book. Factors that should be considered in choosing a method are also discussed.

While bottomhole sampling has the advantage of capturing fluids at reservoir conditions, surface sampling has a potential for obtaining cleaner samples as a result of large volumes of fluid production before sampling.

The choice of either the surface or bottomhole sampling method cannot be considered a simple or routine matter. Each reservoir usually presents certain constraints or circumstances peculiar to it. For example, field operation requirements can impose restrictions on the preparation and execution of a sampling program; sand production or downhole equipment in the well may limit the use of some of the equipment normally used in the sampling operation. Wells that exhibit rapid variations in production rate present special problems in making the necessary measurements with acceptable accuracy. Seasonal or daily weather changes can also influence the sampling operation. Thus, the details of a given sampling procedure often require modification to circumvent local problems. These modifications are usually made based upon on-the-spot judgments. Some of the modifications which can be made to accommodate special situations are presented herein.

Conditioning a well before sampling is almost always necessary. Initial well testing or normal production operations often result in the fluid near the wellbore having a composition which has been altered from that of the original reservoir fluid (for reasons described later). The objective of conditioning the well is to remove this altered (unrepresentative) fluid. Well conditioning consists of producing the well at a rate which will move the altered fluid into the well bore and allow it to be replaced by unaltered (representative) fluid flowing in from further out in the reservoir. Well conditioning is especially important when the reservoir fluid is at or near its saturation pressure at the prevailing reservoir conditions because reduction in pressure near the wellbore, which inevitably occurs from producing the well, will alter the composition of the fluid flowing into the wellbore.

3.3.2 Data Acquisition and QA/QC

The objective of the data-gathering phase is to obtain reliable high-quality data for reservoir evaluation and development. The data requirement depends on the fluid type and the expected development and production strategies (Whitson and Brule 2000). In addition, some heavy oils require customized PVT cells and experimental procedures to accelerate the time needed for attaining equilibrium conditions because of the slow (sometime, nonexistent) gas liberation. On the other hand, more-complex near-critical fluids and miscible-gas-injection processes need special PVT tests and precise measurement techniques to capture the complex phase behavior exhibited by these fluids. Gas condensates in the presence of water require PVT cells that can handle three-phase mixtures of gas, water, and condensate.

A heavy oil sampling program requires extra steps to obtain adequate volumes of representative single-phase oil samples for laboratory analysis. These includes adequate near-wellbore cleaning to minimize sample contamination by drilling-mud filtrate and optimal drawdown to minimize sand production and avoid two-phase flow while mobilizing the oil from the reservoir into the sample chamber (Reddie and Robertson 2004). During surface sampling, measurement uncertainty in the producing GOR is a concern because of large drawdown and incomplete gas separation from the oil. Another issue with surface samples is the slow dissolution of gas while recombining them to prepare reservoir fluid. The main advantage of bottomhole sampling over surface sampling is that the former offers a viable means to capture single-phase samples and eliminate uncertainties associated with surface samples.

The C7+ fraction of the reservoir fluid contains numerous compounds of different homologues (paraffinic, naphthenic, and aromatic) and plays a dominant role in determining the PVT behavior of the fluid. For example, in a gas-condensate fluid, the dew point pressure is (as anticipated) a strong function of the C7+ molecular weight and its relative amount in the fluid. On the other hand, in heavy oils, the C7+ components dictate the viscosity behavior and control the asphaltene-deposition and wax-deposition characteristics of the oil. Similarly, in volatile oils and rich condensates, the oil volumes and other properties below the saturation pressure are determined by the amounts of intermediate and heavy components.

Several methods are used to lump these components into pseudo-components for equation-of-state models (Whitson, 1983). In these models, the C7+ distribution is represented by a continuous gamma distribution that is optimally discretized into a specific number of fractions (i.e., pseudo-components) in which fluid type and the production process involved further guide the component selection.

Because of slow gas liberation and dissolution in heavy oil, special care should be exercised in selecting equipment and procedures for sample preparation and PVT measurements (Cengiz et al. 2004). It is essential to measure the true bubble point pressure as well as the viscosity (for example, by means of a capillary-flow viscometer). Because the oil is saturated at each pressure step in the differential-liberation experiment, small pressure drops in the capillary viscometer caused by the flow will liberate the gas. Therefore, it may be necessary to conduct several viscosity measurements above the saturated pressure and use an extrapolation technique to determine the viscosity at the desired differential-liberation pressure.

PVT-data interpretation and modeling for heavy oils require reliable treatment of C7+ components because a majority of components in heavy oils fall in this range (Ancheyta and Speight, 2007). Solid-forming compounds, such as resin constituents and asphaltene constituents, should be characterized properly for flow assurance needs.

Thus, it is imperative that reservoir fluid characterization studies should relate where possible (considering the nature of heavy oil) to the following issues: (1) acquisition of representative samples at various depths to quantify initial fluid gradients and for PVT studies, (2) PVT measurements to capture near-critical behavior, evaluate gas-injection strategies, and design the surface-separator train, (3) fluid modeling to predict observed near-critical behavior and property changes during gas injection, and (4) development of thermodynamically consistent compositional-gradient models for use in reservoir studies.

Fluid characterization strongly affects in-place-volume, recovery-factor, injectivity/productivity, and well-deliverability calculations. Accurate fluid characterization minimizes technical uncertainties and, thus, provides a reliable representation of the asset value. However, fluid-sampling programs must be tailored to the fluid type, reservoir-rock and reservoir-fluid conditions, and fluid distribution. The fluid type and production processes dictate PVT-data requirements, measurement methods, and data accuracy. In addition, the C7+ components must be characterized accurately. Rigorous modeling methods, such as energy minimization, and robust solution techniques are needed to model near-critical fluids and processes.

Finally, ensuring high-quality data requires routine laboratory visits, evaluation of laboratory procedures and methods, and spot QC as data become available. The QA/QC methods can range from simple graphical techniques to sophisticated material-balance calculations (Whitson and Brule, 2000).

In summary, reliable compositional-gradient models are needed to capture fluid-property variations in reservoirs with high relief and/or near-critical fluids.

3.4 Physical (Bulk) Composition and Molecular Weight

The physical (bulk) composition of reservoir fluids is a subset of fluid characterization (Chapter 4) and distribution within the reservoir that helps in defining reservoir continuity and communication among various zones. Interpretation of well-test data and the design of surface facilities and processing plants require accurate fluid information and its variation with time. In addition to initial reservoir-fluid samples, periodic sampling is necessary for reservoir surveillance.

This book outlines recommended sampling techniques, PVT-data-acquisition strategies, and modeling methods. It also presents field examples covering a wide range of fluid types from heavy oils to lean gas condensates and production processes such as depletion, pressure maintenance, and miscible recovery.

The term *physical composition* (or *bulk composition*) refers to the composition of crude oil as determined by various physical techniques. For example, the separation of petroleum using solvents and adsorbents

(Speight, 2007) into various bulk fractions determines the physical composition of crude oil. These methods of separation are not always related to chemical properties, and the terminology applied to the resulting fractions if often a *terminology of convenience*.

Proper management of production from a heavy oil reservoir can maximize the recovery of the oil originally in the reservoir. Developing proper management strategies requires accurate knowledge of the characteristics of the reservoir fluid, as long as fluid samples obtained from the reservoir fluid reflect the pertinent properties of the fluid, as determined by subsequent laboratory tests.

3.4.1 Sampling

As already stated, the objective of reservoir fluid sampling is to collect a sample that is representative of the fluid present in the reservoir at the time of sampling (Speight, 2001, 2002). The critical steps in any successful sampling program are avoiding two-phase flow in the reservoir, minimizing fluid contamination introduced by drilling and completion fluids, and preserving sample integrity. A sampling program should focus on the key issues of selecting an appropriate sampling method and associated tools; customizing the tool string; and developing sound sampling, sample-transfer, and QC procedures. In addition, specific sampling issues should be addressed related to fluid type and condition, saturation vs. undersaturation, and fluids with non-hydrocarbon components or fluids. Once the sample has been obtained, storage protocols must be observed (Chapter 4). Tests can then commence.

3.4.2 Asphaltene Separation

There are numerous examples of model developments for heavy oil recovery where fractions of the heavy oil are included within the model to incorporate an aspect of heavy oil behavior. The most commonly included fraction is the asphaltene fraction.

However, many models include the asphaltene component that is based upon an average property. This is not only incorrect, but it is dangerous and can lead to incidents. Not only has the asphaltene fraction of the heavy oil been incorrectly separated, but also the average property of the asphaltene fraction in no way represents the true behavior of this complex fraction. The first step to prevent this, of course, is to ensure that the asphaltene fraction is separated by a standard method and remnants of molecular species that belong in other fractions are absent.

On the other hand, heavy oil evaluation by separation into various fractions has been used successfully for several decades. The knowledge of the bulk fractions of heavy oil (Figure 3–5) on a *before-recovery* (core sample analysis) and *after-recovery* (well fluid analysis) basis, as well as variations over time, has been a valuable aid to recovery process development.

The *asphaltene fraction* is that portion of heavy oil feedstock that is precipitated when a large excess (40 volumes) of a low-boiling liquid hydrocarbon (e.g., *n*-pentane or *n*-heptane) is added to the crude oil (1 volume) (Speight, 1994, 2007). *n*-Heptane is the preferred hydrocarbon. *n*-Pentane is still being used in many laboratories, and hexane is used on occasion (Speight 2006, and references cited therein).

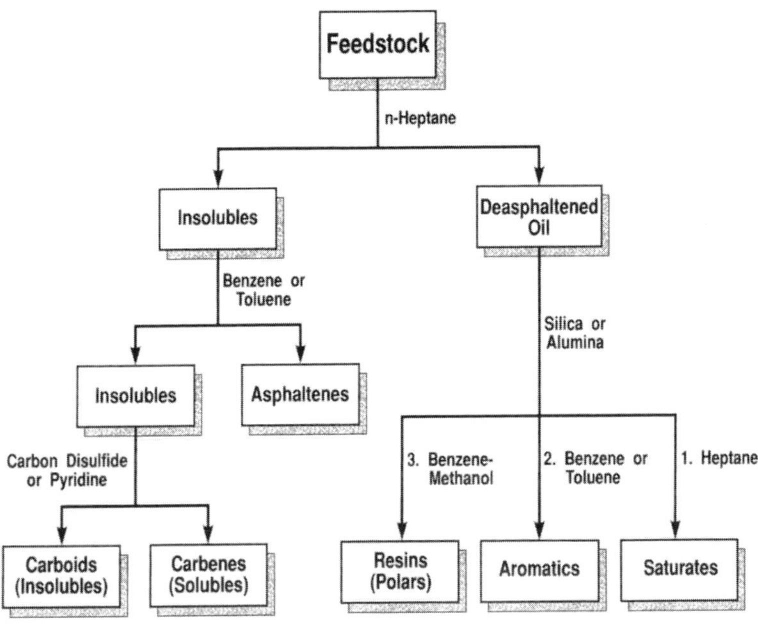

Figure 3–5 *Schematic of the separation of heavy oil into various bulk fractions (see also Figure 2–1).*

Although, *n*-pentane and *n*-heptane are the solvents of choice in the laboratory, other solvents can be used (Speight, 1979). These cause the separation of the asphaltene fraction as a brown-to-black powdery solid material. In the refinery, supercritical low-molecular-weight hydrocarbons (e.g., liquid propane, liquid butane, or mixtures of both) are the solvents of choice, and the product is a semisolid (tacky) to solid asphalt. The amount of asphalt that settles out of the paraffin/residuum mixture depends on the size of the paraffin, the temperature, and the paraffin-to-feedstock ratio (Girdler, 1965; Mitchell and Speight, 1973; Speight et al., 1984).

When pentane and the lower-molecular-weight hydrocarbon solvents are used in large excess, the quantity and the composition of the precipitate changes with *increasing temperature* (Mitchell and Speight, 1973). At ambient temperature, the quantity of precipitate first increases with increasing *ratio of solvent to feedstock* and then reaches a maximum. In fact, for many heavy oils, there are indications that when the proportion of solvent in the mix is less than 35%, little or no asphaltene constituents are precipitated.

Contact time between the hydrocarbon and the feedstock also plays an important role in asphaltene separation. Yields of the asphaltenes reach a maximum after approximately eight hours, which may be ascribed to the time required for the asphaltene particles to agglomerate into particles of a *filterable size* as well as the diffusion-controlled nature of the process. Heavier feedstocks also need time for the hydrocarbon to penetrate their mass.

For example, if the precipitation method (deasphalting) involves the use of solvent and heavy oil, it is essentially a leaching of the soluble constituents from the insoluble constituents and may be referred to as *extraction*. However, under the prevailing conditions now in laboratory use, the term *precipitation* is perhaps more correct and descriptive of the method. Variation of *solvent type* also causes significant changes in asphaltene yield. The contact time between the feedstock and the hydrocarbon liquid can have an important influence on the yield and character of the asphaltene fraction.

At this point, a mention of the phase behavior of asphaltenes and fluids containing asphaltenes is worthwhile. The phase behavior of fluid containing asphaltenes is complex. In general, asphaltenes are difficult to define chemically, whereas the physics and chemistry underlying the definition of this fraction is not open to debate. The fraction is a solubility fraction and is, in reality, an artifact of the separation method (Speight, 2007). In fact, asphaltenes fractions possessing similar constituents may exhibit different properties in their native fluids and in solvents/nonsolvent mixtures. Asphaltenes constituents intra-act and interact with one another and with solvent media (Speight, 2007). Phase behavior and precipitation models must capture the relevant physics and chemistry if derived models are to be truly predictive. But the issue is the use of *average parameters* rather than the recognition that the asphaltene fraction is a collection of different molecular types (Figure 3–6) that vary from crude oil to crude oil (Figure 3–7) (because of the complexities of the maturation process [Speight, 2007]). In fact, analytical methods such as high performance liquid chromatography (HPLC) have shown conclusively that the asphaltene fraction is a mix of unknown (at best, *speculative*) molecular types (Figure 3–8) (Speight, 2001, 2002).

In spite of this, *average parameters* and *average properties* continue to be used for derivation of predictive models. The behavioral characteristics for any average structure for such a complex mixture cannot in any way be truly representative of the behavior of the different constituents. Thus, it is not surprising that many models for heavy oil behavior are of little value in terms of predictive capability, and the models generally flounder (Andersen and Speight, 1999).

3.4.3 Fractionation

Fractionation of heavy oil into components other than asphaltenes (Figure 3–5) has also been of interest in following recovery procedures. By careful selection of a characterization scheme, it may be possible to obtain a detailed overview of oil composition that can be used for process predictions. Thus, fractionation methods also play a role, along with the physical testing methods, in evaluating heavy oils and the various recovery processes, especially when determining whether or not in situ upgrading occurs. For example, by thoughtful selection of an appropriate technique, it is possible to obtain a detailed *map* of heavy oil that can be used for predictions of behavior (Long and Speight, 1989, 1998; Speight, 2007 and references cited therein).

After removal of the asphaltene fraction, further fractionation of petroleum is also possible by variation of the hydrocarbon solvent. For example, liquefied gases, such as propane and butane, precipitate as much as 50% by weight of the residuum or bitumen. The precipitate is a black, tacky, semisolid material, in contrast to the pentane-precipitated

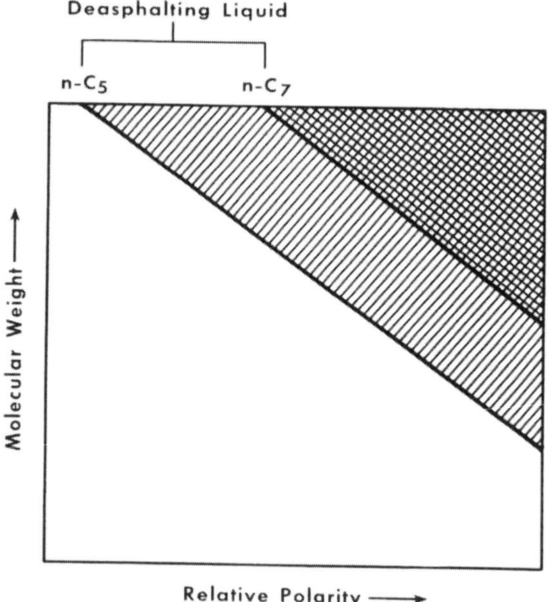

Figure 3–6 *Representation of the asphaltene fraction as a collection of species of different molecular weight and polarity.*

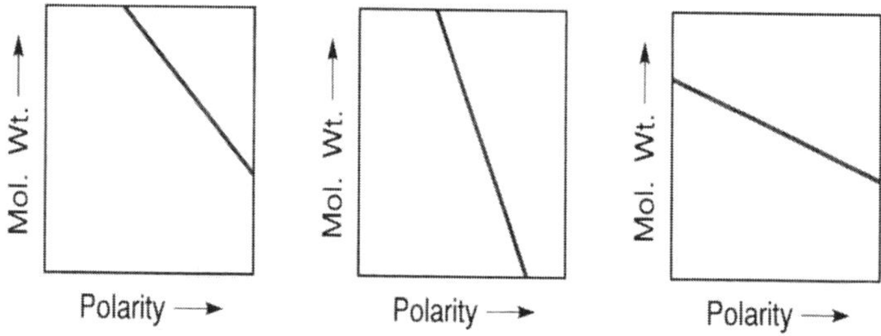

Figure 3–7 *Asphaltenes from different crude oils will vary in the relationship between molecular weight and polarity depending on the relative amounts of the precursors and the maturation process parameters (see Figure 3–5 for comparison).*

asphaltenes, which are usually brown, amorphous solids. Treatment of the propane precipitate with pentane then yields the insoluble brown, amorphous asphaltenes and soluble, near-black, semisolid resins, which are (as near as can be determined) equivalent to the resins isolated by adsorption techniques.

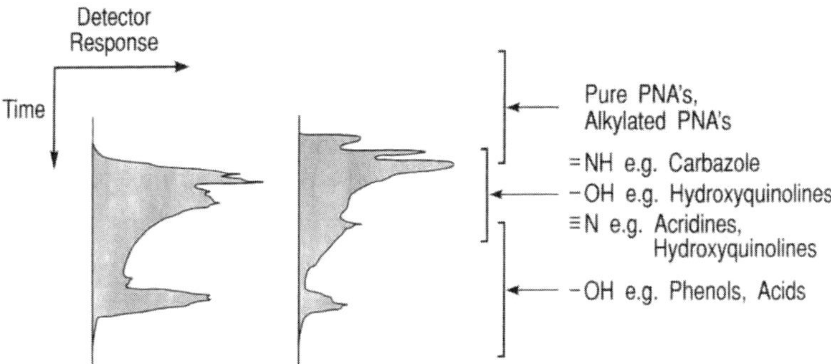

Figure 3–8 *Illustration of the make-up of two different asphaltenes by HPLC.*

Separation by adsorption chromatography essentially commences with the preparation of a porous bed of finely divided solid, the adsorbent. The adsorbent is usually contained in an open tube (column chromatography); the sample is introduced at one end of the adsorbent bed and induced to flow through the bed by means of a suitable solvent. As the sample moves through the bed, the various components are held (adsorbed) to a greater or lesser extent depending on the chemical nature of the component. Thus, those molecules that are strongly adsorbed spend considerable time on the adsorbent surface rather than in the moving (solvent) phase, but components that are slightly adsorbed move through the bed comparatively rapidly.

There are three standard test methods that provide for the separation of heavy oil into four or five constituent fractions (Speight, 2001; ASTM, 2007). It is interesting to note that as the methods have evolved, there has been a change from the use of pentane (ASTM D2006, ASTM D2007) to heptane (ASTM D4124) to separate asphaltenes. This is, in fact, in keeping with the production of a more consistent fraction that represents these higher-molecular-weight, more complex constituents of petroleum (Girdler, 1965; Speight et al., 1984).

Two of the methods (ASTM D2007, ASTM D4124) use adsorbents to fractionate the deasphaltened heavy oil, whereas the third method (ASTM D2006) advocates the use of various grades of sulfuric acid to separate the material into compound types. Caution is advised in the application of this third method since the method does not work well with all feedstocks. For example, when the *sulfuric acid* method

(ASTM D2006) is applied to the separation of heavy oil, complex emulsions can be produced.

Obviously, there are precautions that must be taken when attempting to separate heavy oil into constituent fractions. The disadvantages in using ill-defined adsorbents are that adsorbent performance differs with the same feed and, in certain instances, may even cause chemical and physical modification of the feed constituents. The use of a chemical reactant like sulfuric acid should only be advocated with caution since feeds react differently and the reactant may even cause irreversible chemical changes and/or emulsion formation. These advantages may be of little consequence when it is not, for various reasons, the intention to recover the various product fractions in toto or in the original state. In terms of the compositional evaluation of different feedstocks, however, the disadvantages are very real.

Methods used for the separation of petroleum and heavy oil into various fractions are often identified by the acronyms for the names: PONA (paraffins, olefins, naphthenes, and aromatics), PIONA (paraffins, *iso*-paraffins, olefins, naphthenes, and aromatics), PNA (paraffins, naphthenes, and aromatics), PINA (paraffins, *iso*-paraffins, naphthenes, and aromatics), and SARA (saturates, aromatics, resins, and asphaltenes). It must be recognized that the fractions produced by the use of different adsorbents will differ in content and will also be different from fractions produced by solvent separation techniques. However, for heavy oil fractions, the absence of paraffins in the sample usually precludes many of these acronyms; the most common method is the SARA method.

High-performance liquid chromatography (*HPLC*) has found great utility in separating different hydrocarbon group types and identifying specific constituent types. Of particular interest is the application of the HPLC technique to the identification of the molecular types in the heavy oil, especially the molecular constituents of the asphaltene fraction. The general advantages of HPLC are that: (1) each sample may be analyzed *as received* even though the boiling range may vary over a considerable range and (2) the total time per analysis is usually on the order of minutes.

3.4.4 Molecular Weight

Even though recovery processes, in general, do not affect the quality of the oil, there is still the need to determine the molecular weight of the original constituents as well as the molecular weights of the products to understand the processes. For those original constituents and products that have little or no volatility, e.g., resins and asphaltenes, vapor pressure osmometry (VPO) has been proven to be of considerable value.

There have been numerous attempts made to measure the molecular weight of heavy oil, particularly the asphaltene fraction, using a variety of different methods, but there appears to be a noticeable lack of consensus on the value obtained for a specific sample. This has been irrevocably traced to structural aspects of heavy oil and to the behavior of the asphaltene constituents in molecular dispersion in their own maltenes. In particular, the reliability of the method, the meaning of the term *average molecular weight*, and the interpretation or usefulness of the data have received considerable attention.

Currently, several standard methods are recognized as being useful for determining the molecular weight of petroleum fractions. These methods are

ASTM D2224—Test Method for Mean Molecular Weight of Mineral Insulating Oils by the Cryoscopic Method. (This was discontinued in 1989, but still used by some laboratories for determining the molecular weight of petroleum fractions up to an including gas oil.)

ASTM D2502—Test Method for Estimation of Molecular Weight (Relative Molecular Mass) of Petroleum Oils from Viscosity Measurements.

ASTM D2503—Test Method for Estimation of Molecular Weight (Relative Molecular Mass) of Hydrocarbons by Thermoelectric Measurement of Vapor Pressure.

ASTM D2878—Method for Estimating Apparent Vapor Pressures and Molecular Weights of Lubricating Oils.

Methods for molecular weight measurement are also included in other more comprehensive standards (ASTM D128, ASTM D3712) and there are several indirect methods have been proposed for the estimation of molecular weight by correlation with other, more readily measured, physical properties. They are satisfactory when dealing with the conventional type of crude oils or their fractions and products and when approximate values are desired.

Vapor pressure osmometry (ASTM D2503), also called vapor phase osmometry, is a relatively simple and cheap method for the determination of molecular weight. Most osmometers can operate over a range of temperature through the use of probes that cover specific temperature ranges. This gives the *number average molecular weight* and not the molecular weight distribution.

The measuring elements are two temperature-sensitive thermistors placed in a closed, heat-insulated chamber. By means of two syringes, a drop of solvent is placed on one and a drop of the sample in solution in the same solvent is placed on the other. The chamber is saturated with solvent vapor and is carefully temperature controlled. Because the solution has a lower vapor pressure than the solvent, solvent from the chamber atmosphere will condense on it. The difference in vapor pressure is proportional to its sample concentration. The heat of condensation warms the solution drop until its vapor pressure is the same as that of the surroundings. From then on, a steady state of condensation and warming is established. The temperature increase is measured and recorded. The solvent drop on the other thermistor is in equilibrium with the solvent in the chamber and, theoretically, gives no temperature effect. In reality, convection and other effects cause minute disturbances in the system, which can be minimized by subtracting the voltage of the solvent thermistor from that of the solution thermistor.

The temperature difference ΔT, is related to concentration and molecular weight at infinite dilution by the equation

$$\Delta T = \mathrm{K}_1 c/MW$$

where K_1 is a constant determined by calibration and c is the solute concentration. The effect is measured at several concentrations, and the results are plotted versus the reciprocal of the molecular weight ($1/MW$) and extrapolated to zero concentration (also referred to as unlimited dilution). However, one of the issues that arises through the use of this method for determining the number average molecular weight is that there can be a shift to the low values by the presence of low-molecular-weight contaminants, as might be used in preparing heavy oil for transport.

A common solvent for vapor pressure osmometry is toluene, which is satisfactory for hydrocarbons and moderately polar compounds. However, for the highly polar fractions, such as asphaltenes, more polar solvents such as pyridine are required. The molecular weight of such fractions measured by vapor pressure osmometry in pyridine are distinctly lower than those measured in toluene (Speight, 2001 and references cited therein) indicating a lower degree of aggregation, assuming that contamination with trace amounts of previously used solvent in the separation of the samples can truly be excluded.

The observance of lower-than-true molecular weights due to the influence of solvent and/or adsorbed lower-molecular-weight species on the asphaltenes' constituents is not specific to the vapor pressure osmometry method. It applies to any method producing number average results. Only further fractionation or other suitable measurements can provide the molecular weight distribution of these polar fractions. Re-precipitation or further fractionation of asphaltenes mitigates the preferential solvation of large polar molecules by small ones (Speight, 2001 and references cited therein).

One technique for reducing the effect of aggregation is to measure the molecular weight of polar species (asphaltenes) at different concentrations and different temperatures to illustrate and negate any aggregation effects. The molecular weight is measured at three different concentrations at each of three different temperatures (Speight, 1994). The data are extrapolated to determine the molecular weight at zero concentration at each temperature and then extrapolated to room temperature to negate concentration and temperature effects. This, of course, assumes that the extrapolation line is straight.

The *aggregation effect* has, on occasion, been incorrectly referred to as an *error*, is that the accuracy of the molecular weight value is influenced by *aggregation* of the molecular species in solution. In some instances, this may not be classed as an error, as long as the aggregation is recognized and the method is used (along with molecular weight determination in a polar solvent) to determine the maximum aggregation that can occur in the nonpolar solvent. Such effects are magnified when measuring the molecular weight of the asphaltenes that consist primarily of polar (heteroatom) constituents. Indeed, from such experiments, the aggregation effect can be determined as the ratio of the molecular weight in a nonpolar solvent (e.g., toluene) to the molecular weight in a polar solvent (e.g., pyridine or nitrobenzene). Thus,

$$\text{Aggregation factor } N = MW_{\text{toluene}}/MW_{\text{pyridine}}$$

$$\text{Aggregation factor } N = MW_{\text{toluene}}/MW_{\text{nitrobenzene}}$$

The method using the concept of *freezing point depression* to determine the average molecular weight of various fractions uses simple equipment. The samples must be completely soluble and chemically inert in the solvent (benzene). Molecular weight values represent a number average molecular weight for the sample components.

The freezing points for benzene and for a benzene solution containing a known weight of sample are determined from their cooling curves. From the depression of the freezing point due to the sample, the average molecular weight is calculated by application of Raoult's law, the vapor pressure lowering of a solvent is directly proportional to the concentration of the solute:

$$(p_o - p)/p_o = n_2/(n_1 + n_2)$$

Where n_1 and n_2 are the number of moles of solvent and solute, respectively. Thus, if w_2 grams of solute of molecular weight M_2 is dissolved in w_1 grams of solvent of molecular weight M_1, Raoult's law gives

$$(p_o - p)/p_o = (w_2/M_2)/(w_1M_1 + w_2M_2)$$

For a dilute solution, n_2 may be neglected in favor of n_1. Thus, the calculation for molecular weight determination by cryoscopy is

$$M = \{W(K_f - \Delta T)]/(\Delta T.S)$$

where M is the number average molecular weight, W is the weight of sample in grams, K_f is the cryoscopic constant for the solvent (K_f for benzene = 65.6), ΔT is the freezing point depression in degrees Celsius (i.e., the freezing point of benzene minus the freezing point of the solution), and S is the moles of solvent. Many solvents have been tried, but with the best technique, an accuracy of 1% to 2% is considered good.

Boiling point elevation (*ebullioscopic*) methods are, in general use, more rapid and equally accurate, but they tend to fall short when applied to the higher-molecular-weight fractions of petroleum, heavy oil, and bitumen. Molecular weights in the low range (<500 daltons) are readily determined by vapor-density methods.

For any one sample, the boiling point elevation is determined at a series of concentrations of solute. Such determinations are carried out in practice by comparing the boiling point of pure solvent, measured in an ebulliometer, with the boiling points of a series of solutions prepared by adding successive portions of the sample to the solvent, determined in a second ebulliometer. Measurement with the first ebulliometer serves as the control experiment.

In order to determine the ebullioscopic constant K—which depends not only on the solvent used, but also to a certain extent on the construction of the apparatus and the procedure followed—a series of determinations is carried out on a pure hydrocarbon of known molecular weight, preferably of the same order of magnitude as the molecular weight under investigation. The ebullioscopic constant calculated for each concentration is plotted on a graph, just as in the case of a molecular weight determination, and is extrapolated to zero concentration.

Before attempting to measure molecular weights, care should be taken to ensure that the oil sample does not contain water or volatile components (boiling below 200°C, 392°F) because this would invalidate the basic calculation.

Boiling point elevation methods all give number average molecular weights, defined by

$$\Sigma(n_i M_i)/\Sigma n_i = M_n$$

where M_n is the number average molecular weight, and n_i is the number of molecules having a molecular weight M_i.

Size exclusion (gel permeation) chromatography (ASTM D5296) methods can determine the molecular weight of petroleum, heavy oil, bitumen and their constituent fraction. The principle of size exclusion chromatography is the exclusion of larger sample molecules from smaller pores in the packing. As a result, larger molecules cannot reside in the entire column volume but instead are restricted to smaller regions. In the extreme, the largest ones are restricted to the interstitial volume, i.e., the. space between particles, whereas the smallest ones can penetrate the entire open column volume, i.e., the interstitial and all the pore volume. As a consequence, the large molecules elute first and the smallest ones last.

For the higher-molecular-weight petroleum fractions, such as the asphaltene fraction of heavy oil, column packing of nominal pore sizes between 50 to 100 Å and 5,000 to 10,000 Å is used for complete resolution. The bimodal column combinations give good linear calibration curves and well-characterized polymer samples. Narrow polymer fractions, e.g., polystyrenes, are available for calibrating column sets across the entire molecular weight range.

However, these polymers are structurally different from petroleum constituents and cannot be directly used for accurate calibration. In fact, petroleum fractions, such as the resin fraction and the asphaltene fraction, differ from polymers in three ways: (1) the hydrocarbon skeletal structure, (2) the polar nature, and (3) the varying composition with molecular weight. Therefore, it is preferable to employ a set of *narrow* subfractions obtained by preparative size exclusion chromatography from the same or a similar petroleum fraction for calibration. Calibration curves obtained by polymers should at least be checked with such subfractions and adjusted as necessary.

It is important to remember that size exclusion chromatography separates by molar volume rather than by molecular weight. The method will, therefore, differentiate by structure in addition to molecular weight. In principle, size exclusion chromatography is a very powerful method for separating petroleum fractions by molecular weight. It is used frequently in petroleum analysis despite the components toward adsorption and aggregation and other potential problems.

3.5 Reservoir Evaluation

It has been customary to attempt to define conditions—usually a limited list of reservoir and fluid properties and operating conditions—under which thermal recovery operations are likely to be commercially attractive. These conditions have been called by a variety of names, including screening criteria, screening guidelines, preferred criteria, and selection criteria. Any such set of criteria reflects its authors' opinions at a particular time and usually are affected by the current and local economic climate.

Although most available guidelines attempt to include the effect of economic considerations in the technical variables, there are few technical considerations that in themselves limit the operability of thermal recovery processes. Important technical considerations are those that affect: (1) the ability to generate heat within or inject heat into an oil-containing reservoir at efficient rates, (2) the ability to displace the oil, and (3) the ability to recover the oil, all in a controlled manner. It is difficult, if not impossible, to translate these broad considerations into specific values of properties or groups of properties that would limit the range of applicability of a thermal process. Some examples will help clarify this point.

In steam drives, the gravity of the crude oil plays no role in the technical considerations just listed except as it might affect plugging and, thus, the ability to maintain adequate communication between wells in reservoirs containing relatively heavy crude oils. With regard to loss in injectivity or productivity, gravity override (or bypassing) of steam reduces the tendency of the formation to plug. Also, hydraulic fracturing, control of the injection temperature, and cyclic steam injection have been used successfully in thermal operations to avoid or minimize plugging. Thus, it seems impractical to place a limit on the range of API gravity of crude oils to be considered for steam drive processes.

As another example, consider injectivity. Reduced ability to inject (or produce) fluids is a factor affecting the economics of all enhanced recovery processes, especially those requiring the injection of heated fluids. When the injection rate is reduced, the project life must be longer, and more of the heat entering the formation is lost to the adjacent zones. This tends to reduce the oil displacement and production rates (i.e., the rate at which income is generated), which in turn reduces the economic attractiveness. But what may be an acceptable level of injectivity in one project could be economically disastrous in another.

Although guidelines giving parameter limits are useful in helping government bodies and oil companies decide how to allocate funds to support their general research and development efforts, they are of little use to an operator having a specific reservoir requiring either improved performance or special development. Each reservoir should be examined individually (at least briefly) as though there were no guidelines, especially where the reserves are great enough to support some engineering studies. A screening economic analysis for each likely process is the preferred method for forming initial decisions.

The approach to this analysis is to indicate some of the elements necessary to screen a potential venture, including an understanding of the possible effects of uncertainties on a parameter or group of parameters. This approach, which requires only a few hours to perform, should allow the engineer to make preliminary decisions or recommendations as to the merits of a prospective thermal project.

Reservoir thickness is particularly important because it has a pronounced effect on the fraction of the introduced heat that remains in the reservoir, a quantity known as the heat efficiency. For a given reservoir (which has a fixed thickness), the only way to increase the heat efficiency is to reduce the time required to produce the oil from the zone between an injector and a producer. This generally can be done by increasing the injection rate (one also must be able to recover the oil at an increased rate) or by reducing the spacing. Each of these alternatives would cost money to implement but may be attractive compared with a slowly expanding steam front that results in low oil production rates and a long operating life.

The main consideration regarding well spacing is sand continuity, the determination of which may require the combined interpretation of geologists, stratigraphers, and logging and petrophysical experts. Sand continuity is vital in the selection of well spacing for processes involving interwell fluid displacement. Partial communication in sands where only a fraction of the interval correlates between well locations invariably leads to poor recoveries and poor effective transmissivity. Knowledge of the depositional setting of the sands may be crucial in establishing whether open intervals in nearby wells are part of the same blanket sand or are likely to be in unconnected and meandering river channels. Regional trends in the direction of river channels may result in an apparent anisotropy in reservoir permeability, with relatively good communication between wells in the same channels and poor or no communication in a transverse direction.

Displacement processes are not indicated where communication is poor over relatively short distances. Instead, cyclic steam injection or some other stimulation treatment that does not depend on interwell communication is preferred.

Another factor related to well spacing is the well pattern. Because of the extensive use of repeated well patterns (such as five-spots) in waterflooding, they were considered necessary in the early years of thermal recovery. Now, however, there is considerable evidence that the use of repeated, regular well patterns is not always essential for the success of field projects. The choice of well locations is important in reservoirs having significant dip or a strong lateral aquifer influx.

In addition to the factors already discussed, there are a few other important considerations in selecting a reservoir for thermal projects. Foremost are any factors that may affect the control of the operation, especially where safety is involved.

Control means that the fluids are injected only into the target reservoir and that both displaced and injected fluids are collected efficiently and safely. The leakage of produced or injected fluids to the surface is an extreme example of loss of control.

Conditions that should be considered and evaluated carefully are injection pressures close to the fracture initiation pressure or close to the estimated pressure limits of well equipment (especially in old wells) and the likelihood of communication with nearby outcrops. These could lead to fluid contamination of overlying formations, loss of production, loss of pressure, loss of control, and possibly both unexpected workover costs and early termination of the project.

Where displaced oil can bypass a producing well, it is generally desirable on technical grounds (but not necessarily justifiable on economic grounds) to fracture the well and to maintain it in unimpaired condition, especially where low-flow-resistance zones near the well compete for the flow of oil. Stimulation of producers by cyclic steam injection has long been recognized as an excellent method for reducing the resistance to flow into these wells. Special attention should be paid to methods for avoiding or minimizing the resaturation of previously depleted zones or the saturation of previously clean sands (such as zones below a water/oil contact).

3.6 References

Ancheyta, J., and Speight, J.G. 2007. Hydroprocessing of Heavy Oils and Residua. CRC-Taylor and Francis Group, Boca Raton, Florida.

Andersen, S.I., and Speight, J.G. 1999. Thermodynamic Models for Asphaltene Solubility and Precipitation. *Journal of Petroleum Science and Engineering* 22:53.

Anderson, W.G., 1986. Wettability literature survey Part1. Rock-oil-brine interactions and the effects of core handling on wettability. *Journal of Petroleum Technology* 38:1125–1144.

API. 2003. Sampling Petroleum Reservoir Fluids. Report API RP 44. American Petroleum Institute, Washington, DC.

ASTM. 2007. Annual Book of Standards. American Society for Testing and Materials West Conshohocken, Pennsylvania.

ASTM D128-98(2003) Standard Test Methods for Analysis of Lubricating Grease. Annual Book of Standards. American Society for Testing and Materials, West Conshohocken, Pennsylvania.

ASTM D2006. Standard Test Method for Characteristic Groups in Rubber Extender and Processing Oils by the Precipitation Method (Withdrawn 1975). Annual Book of Standards. American Society for Testing and Materials, West Conshohocken, Pennsylvania.

ASTM D2007. Standard Test Method for Characteristic Groups in Rubber Extender and Processing Oils and Other Petroleum-Derived Oils by the Clay-Gel Absorption Chromatographic Method. Annual Book of Standards. American Society for Testing and Materials, West Conshohocken, Pennsylvania.

ASTM D2503. Standard Test Method for Relative Molecular Mass (Molecular Weight) of Hydrocarbons by Thermoelectric Measurement of Vapor Pressure. Annual Book of Standards. American Society for Testing and Materials, West Conshohocken, Pennsylvania.

ASTM D3712. Standard Test Method of Analysis of Oil-Soluble Petroleum Sulfonates by Liquid Chromatography. Annual Book of Standards. American Society for Testing and Materials, West Conshohocken, Pennsylvania.

ASTM D4124. Standard Test Methods for Separation of Asphalt into Four Fractions. Annual Book of Standards. American Society for Testing and Materials, West Conshohocken, Pennsylvania.

ASTM D5296. Standard Test Method for Molecular Weight Averages and Molecular Weight Distribution of Polystyrene by High Performance Size-Exclusion Chromatography. Annual Book of Standards. American Society for Testing and Materials, West Conshohocken, Pennsylvania.

Caruana, A., and Dawe, R.A. 1996a. Effect of Heterogeneities on Miscible and Immiscible Flow Processes in Porous Media, Trends in Chemical Engineering. 3:185–203.

Caruana, A., and Dawe, R.A. 1996b. Flow Behaviour in the Presence of Wettability Heterogeneities. Transport in Porous Media. 25:217–233.

Cengiz, S., Robertson, C., Kalpacki, B., and Gupta, D. 2004. A Study of Heavy Oil Solution Gas Drive for Hamaca Field Depletion Studies and Interpretations. Paper SPE 86967-MS. SPE International Thermal Operations and Heavy Oil Symposium and Western Regional Meeting, Bakersfield, California, March 16–18.

Dawe R.A. 2003. Reservoir Engineering, 2000. In Modern Petroleum Technology, Upstream Volume, ed. R.A.Dawe, 207–282. Institute of Petroleum, John Wiley and Sons, Chichester, England.

Dawe, R.A. 2004. Miscible Displacement in Heterogeneous Porous Media. Proceedings. Sixth Caribbean Congress Of Fluid Dynamics. Uwi, January 22–23.

Girdler, R.B. 1965. Proceedings of the Association of Asphalt Paving Technologists. 34:45.

Grattoni, C.A., and Dawe, R.A. 2003. Consideration of Wetting and Spreading in Three-Phase Flow in Porous Media, Progress in Mining and Oilfield Chemistry, Volume 5. Recent Advances in Enhanced Oil and Gas Recovery, ed. I. Lakatos. Akad. Kiado, Budapest.

Honarpour, M.M., Nagarajan, N.R., and Sampath, K. 2006. Rock/Fluid Characterization and Their Integration—Implication on Reservoir Management. Journal of Petroleum Technology. 58(9):120.

Kovscek, A.R. 2002. Heavy and Thermal Oil Recovery Production Mechanisms. Quarterly Technical Progress Report, Reporting Period April 1 through June 30, 2002. DOE Contract Number: DE-FC26-00BC15311. United States Department of Energy, Washington, DC

Long, R.B., and Speight, J.G. 1989. Studies in Petroleum Composition. I Development of a Compositional Map for Various Feedstocks. Revue de l'Institut Français du Pétrole. 44:205.

Long, R.B., and Speight, J.G. 1998. The Composition of Petroleum. In Petroleum Chemistry and Refining, ed. J.G. Speight. Taylor & Francis, Washington, DC.

Mitchell, D.L., and Speight. J.G. 1973. The Solubility of Asphaltenes in Hydrocarbon Solvents. Fuel. 52:149.

Nagarajan, N.R., Honarpour, M.M., and Sampath, K. 2007. Reservoir-Fluid Sampling and Characterization—Key to Efficient Reservoir Management. Paper SPE 103501 and 101517. 2006 Abu Dhabi International Petroleum Exhibition & Conference, Abu Dhabi, November 5–8.

Reddie, D.R. and Robertson, C.R. 2004. Innovative Reservoir Fluid Sampling Systems. Paper SPE 86951-MS. SPE International Thermal Operations and Heavy Oil Symposium and Western Regional Meeting, Bakersfield, California. March 16–18.

Speight, J.G. 1979. Information Series No. 84. Alberta Research Council, Edmonton, Alberta, Canada.

Speight, J.G., Long, R.B., Trowbridge, T.D. 1984. Factors Influencing the Separation of Asphaltenes from Heavy Petroleum Feedstocks. Fuel. 63:616.

Speight, J.G., Wernick, D.L., Gould, K.A., Overfield, R.E., Rao, B.M.L., and Savage, D.W. 1985. Molecular Weights and Association of Asphaltenes: A Critical Review. Revue de l'Institut Français du Pétrole. 40:27.

Speight, J.G. 1994. Chemical and Physical Studies of Petroleum Asphaltenes. In Asphaltenes and Asphalts, I. Developments in Petroleum Science, 40. ed. T.F. Yen and G.V. Chilingar, Chapter 2. Elsevier, Amsterdam, The Netherlands.

Speight, J.G. 2001. Handbook of Petroleum Analysis. John Wiley & Sons Inc., New York.

Speight, J.G. 2002. Handbook of Petroleum Product Analysis. John Wiley & Sons Inc., Hoboken, New Jersey.

Speight, J.G. 2007. The Chemistry and Technology of Petroleum. 4th Edition. CRC-Taylor & Francis Group, Boca Raton, Florida.

Whitson, C.H. 1983. Characterizing Hydrocarbon Plus Fractions. Society of Petroleum Engineers Journal. 23(4):683.

Whitson, C.H., and Brule, M. 2000. Phase Behavior. Monograph Series. Society for Petroleum Engineers, Richardson, Texas.

Witt, C.J., and Crombie, A. 1999. A Comparison of Wireline and Drillstem Test Fluid Samples From a Deep Water Gas-Condensate Exploration Well. Paper SPE 56714-MS. SPE Annual Technical Conference and Exhibition, Houston, October 3–6.

CHAPTER 4

PROPERTIES

Petroleum is perhaps the most important substance consumed in modern society. It provides not only raw materials for the ubiquitous plastics and other products, but also fuel for energy, industry, heating, and transportation. From a chemical standpoint, petroleum is an extremely complex mixture of hydrocarbon compounds, with minor amounts of nitrogen-, oxygen-, and sulfur-containing compounds as well as trace amounts of metal-containing compounds.

The fuels that are derived from petroleum supply more than half of the world's total supply of energy. Gasoline, kerosene, and diesel oil provide fuel for automobiles, tractors, trucks, aircraft, and ships. Fuel oil and natural gas are used to heat homes and commercial buildings, as well as to generate electricity. Petroleum products are the basic materials used for the manufacture of synthetic fibers for clothing and in plastics, paints, fertilizers, insecticides, soaps, and synthetic rubber. The uses of petroleum as a source of raw material in manufacturing are central to the functioning of modern industry. As petroleum resources are depleted, industry will rely more and more on heavy oil to satisfy the need for fuels, chemicals, and the other products currently derived from petroleum.

Many types of heavy oil exist, and a variety of production processes are being used and developed to recover it. However, technologies and services used for conventional oil face limitations with heavy oil.

Basically, there is no direct positive correlation between the *heaviness* of heavy oil and the extent of paraffin and/or asphaltene problems associated with the production operation. In fact, there are many producers of heavy oil that do not experience severe paraffin or asphaltene problems, while there are numerous production operations that experience rather severe paraffin or asphaltene problems even though they are not producing heavy oil.

Heavy oil typically has relatively low proportions of volatile compounds with low molecular weights and quite high proportions of high molecular weight compounds. The high-molecular-weight fraction of heavy oils is composed of compounds (not necessarily paraffins or asphaltenes) with high melting points and high pour points that greatly contribute to the fluid properties of heavy oil and hence to reduced mobility compared to conventional petroleum. It is typically this poor mobility of the crude oil, as opposed to accumulations of paraffins or asphaltenes in formation rock pore throats or production lines, that is usually the cause of production problems.

Some, but not all, heavy oils do contain moderate-to-high levels of asphaltene constituents. However, the asphaltene constituents do not become a problem unless they drop out of solution (precipitate) and build up in the formation or production string.

In summary, the heaviness of heavy oil is primarily the result of an internal balance between a relatively high proportion of complex, high-molecular-weight, non-paraffinic compounds and a low proportion of volatile, lower-molecular-weight compounds. The problems of producing heavy oil from the reservoir are typically a result of disturbing the internal balance, which, in turn, influences the mobility of the oil and the deposition of asphaltene constituents. Success with heavy oil depends as much on understanding the fluid properties of the reservoir as it does on knowing the geology of the reservoir itself. The reason is that the chemical differences between heavy oil and conventional oil ultimately affect their viscosity.

Thermal recovery options in some reservoirs include the use of cyclical steam (*huff 'n' puff*), downhole heaters, or a relatively new commercial process called steam-assisted gravity drainage (SAGD). Other techniques, such as injecting slugs of water alternating with gas (WAG) are less efficient than thermal recovery but also less expensive. The use of steam influences the mobility of the oil, which influences recovery rates, but the enhanced oil recovery and artificial lift methods needed to produce changes to the already complex fluid characteristics of heavy oil.

Strategies for recovering heavy oil emphasize the difference in properties that influences the choice of methods for conversion to various products, and the availability of processes that can be employed to produce heavy oil from the reservoir has increased significantly in recent years (Speight, 2007). In order to determine the recoverability of heavy oil, a series of consistent and standardized characterization procedures are required (ASTM, 2007). These procedures can be used with a wide variety of feedstocks to develop a general approach to predict behavior during recovery (Speight, 2001).

The recovery of heavy oil can be designed in an optimal manner by performing selected evaluations that lead to an understanding of the chemical and physical features of the oil. The evaluation schemes do not need to be complex but must focus on key parameters that affect recovery. For example, the identification of the important features can be made with a saturates-aromatics-resins-asphaltene (SARA) separation. Subsequent analysis of the fractions also provides further information about the intramolecular relationships of the oil. In this follow-on analysis, there is often emphasis on the asphaltene fraction since the solubility of asphaltene constituents and the thermal products has a dramatic affect on solids deposition during recovery. Further evaluation of the asphaltene constituents can lead to correlation of asphaltene properties with behavior. In addition, useful information can be gained from knowledge of elemental composition and molecular weight as well as size exclusion chromatography and high performance liquid chromatography profiles.

4.1 Physical Properties

Heavy oil exhibits a wide range of physical properties, and several relationships can be made between various physical properties (Speight, 2007). Although the properties such as viscosity, density, and boiling range may vary widely, the ultimate or elemental analysis varies over a narrow range for a large number of samples. The carbon content is relatively constant, whereas the hydrogen and heteroatom contents are responsible for the major differences between heavy oils.

Heavy oil containing 9.5% heteroatoms may contain few pure hydrocarbon constituents; most constituents contain *at least one or more* nitrogen, oxygen, and/or sulfur atoms within the molecular structures. It is the presence of these heteroelements that can have substantial effects on the recovery process. Coupling this with the changes brought about to the feedstock constituents by disturbing the delicate intramolecular balance, it is not surprising that recovering heavy oil can be a monumental task.

Initial inspection of the oil (conventional examination of the physical properties) is necessary. From this, it is possible to make deductions about the propensity for easy or difficult recovery. In fact, evaluation of heavy oil from physical property data to determine which recovery sequences should be employed for a particular crude oil is a predominant part of the initial examination of any heavy oil. *Proper* interpretation of the data resulting from the inspection of crude oil requires an understanding of their significance (Speight, 2001, 2007).

The chemical composition of heavy oil, however, is a much truer indicator of behavior than its physical properties. Whether the composition is represented in terms of compound types or (more likely) in terms of generic compound classes, it can assist in determining the nature of any potential interactions of the oil with the rock, for example, or with changes in pressure and temperature. Hence, chemical composition can play a large part in determining the nature of the products that arise from the recovery operations. It can also play a role in determining the means by which a particular feedstock should be processed (Speight, 2007). This becomes particularly important when partial upgrading in the reservoir is considered as a serious option for recovery.

To satisfy specific needs with regard to heavy oil recovery, as well as to the nature of the recovered product, various standard test methods are available from organizations such as the American Society for Testing and Materials (ASTM, 2007). A complete discussion of the large number of routine tests available for petroleum, heavy oil, and petroleum residua fills an entire book (ASTM, 2007). It seems appropriate that in any discussion of the physical properties of heavy oil reference should be made to the corresponding test; accordingly, the various test numbers have been included in this text. Some of the tests have been abandoned but are still used in many laboratories, especially where a replacement test has not been developed. These test methods are also included here.

Before any analysis of test results occurs, it is necessary to ensure that the sample is consistent with the reservoir and that the data can be reproduced.

4.1.1 Sampling

Because of the complexity of heavy oil, the importance of the correct *sampling* of heavy oil (Wallace, 1988; Speight, 2000) cannot be over-stressed. Properties such as elemental analysis, metals content, density (specific gravity), and viscosity are affected by the homogeneity (or heterogeneity) of the sample. In addition, adequate records of the circumstances and conditions during sampling have to be made. For example, in sampling from oil-field separators, the temperatures and pressures of the separation plant and the atmospheric temperature would be noted. An accurate sample handling and storage log should be maintained and should include information such as:

1. The precise source of the sample, i.e., the exact geographic location or refinery locale from which the sample was obtained
2. A description of the means by which the sample was obtained
3. The protocols used to store the sample
4. Chemical analyses, such as elemental composition
5. Physical property analyses, such as API gravity, pour point, distillation profile, etc.
6. ASTM methods used to determine the properties in items 4 and 5
7. The number of times that the sample has been retrieved from storage to extract a portion, i.e,. indications of exposure to the air or oxygen

Attention to factors such as these enables standardized comparisons to be made when subsequent samples are taken.

There are several protocols involved in the initial isolation and clean-up of the sample. In fact, considerable importance attaches to the presence of *water* or *sediment* in crude oil (ASTM D1796, ASTM D4007) because they lead to difficulties in the refinery, e.g., corrosion of equipment, uneven running on the distillation unit, blockages in heat exchangers, and adverse effects on product quality.

The sediment consists of finely divided solids that may be dispersed in the oil or carried in water droplets. The solids may be drilling mud, sand, or scale picked up during the transport of the oil, or they may consist of chlorides derived from evaporation of brine droplets in the oil. In any event, the sediment can lead to serious plugging of the equipment, corrosion due to chloride decomposition, and a lowering of residual fuel quality.

Water may be found in the crude oil either in an emulsified form or in large droplets. It can cause flooding of distillation units and excessive accumulation of sludge in tanks. The quantity is generally limited by pipeline companies and by refiners, and steps are normally taken at the wellhead to reduce the water content as much as possible. However, water can be introduced during shipment. In any form, water and sediment are highly undesirable in a refinery feedstock, and the relevant tests (ASTM D954, ASTM D1796, ASTM D4006) are regarded as important in crude oil quality examinations. Prior to assay, it is sometimes necessary to separate the water from a crude oil sample. This is usually carried out by one of the procedures described in the preliminary distillation of crude oil (IP 74).

Careful sampling of a crude oil is necessary when examining for suspended impurities of this type.

There is a great variation in the *salt content* of crude oil, depending mainly on the source and possibly on the producing wells or zones within a field. In addition, at the refinery, salt water introduced during shipment by tanker may contribute to this total salt content. These salts have adverse effects on refinery operations, especially in increasing maintenance following corrosion in crude units and heat exchangers. It is common practice to monitor wells in a producing field for high salt content, and it is also general practice to desalt the crude oil at the refinery. As with water and sediment tests, careful sampling is necessary when determining the salt content of crude oil. It would appear that further tests to determine the corrosiveness due to the individual chemical components of salt and to determine the extent of evolution of hydrogen chloride on heating would be desirable.

4.1.2 Elemental (Ultimate) Analysis

The analysis of heavy oil for the percentages of carbon, hydrogen, nitrogen, oxygen, and sulfur is perhaps the first method used to examine and evaluate the general nature of a feedstock. The atomic ratios of the various elements to carbon (i.e., H/C, N/C, O/C, and S/C) are frequently used for indications of the overall character of the heavy oil. It is also of value to determine the amounts of trace elements, such as vanadium and nickel, in a feedstock since these materials can have serious deleterious effects on catalyst performance during partial upgrading during recovery or even when using a partial upgrading process at the surface before transportation.

There are procedures (ASTM, 2007) for the ultimate analysis of petroleum and petroleum products, but many such methods may have been designed for other materials. For example, *carbon content* can be determined by the method designated for coal and coke (ASTM D3178) or by the method designated for municipal solid waste (ASTM E-777). There are also methods designated for:

1. *hydrogen content* (ASTM D1018, ASTM D3178, ASTM D3343, ASTM D3701, and ASTM E-777);
2. *nitrogen content* (ASTM D3179, ASTM D3228, ASTM D3431, ASTM E-148, ASTM E-258, and ASTM E-778);
3. *oxygen content* (ASTM E-385)
4. *sulfur content* (ASTM D124, ASTM D1266, ASTM D1552, ASTM D1757, ASTM D2662, ASTM D3177, ASTM D4045 and ASTM D4294).

Of the data that are available, the proportions of the elements in heavy oil vary only slightly over narrow limits. Perhaps the more pertinent property in the present context is the sulfur content; sulfur content and API gravity represent the two properties that have the greatest influence on the value of heavy oil. The sulfur content varies from about 0.1% to about 5% by weight (Speight, 2007 and references cited therein).

4.1.3 Metals Content

Metals (particularly vanadium and nickel) are found in most crude oils (Reynolds, 1998). Heavy oil contains relatively high proportions of metals, either in the form of salts or as organometallic constituents (such as the metalloporphyrins), which are extremely difficult to remove from the feedstock. The metallic constituents may actually *volatilize* under thermal recovery operations and appear in the reservoir or in the production lines.

A variety of tests have been designated for the determination of metals in heavy oil, petroleum, and petroleum products (ASTM D1026, ASTM D1262, ASTM D1318, ASTM D1368, ASTM D1548, ASTM D1549, ASTM D2547, ASTM D2599, ASTM D2788, ASTM D3340, ASTM D3341, and ASTM D3605). Determination of metals in whole feeds can be accomplished by combustion of the sample so that only inorganic ash remains. The ash can then be digested with an acid and the solution examined for metal species by atomic absorption (AA) spectroscopy or by inductively coupled argon plasma (ICAP) spectrometry.

4.1.4 Density and Specific Gravity

Density is the mass of a unit volume of material at a specified temperature; it has the dimensions of grams per cubic centimeter (a close approximation to grams per milliliter). *Specific gravity* is the ratio of the mass of a volume of the substance to the mass of the same volume of water and is dependent on two temperatures, those at which the masses of the sample and the water are measured. When the water temperature is 4°C (39°F), the specific gravity is equal to the density in the centimeter-gram-second (cgs) system since the volume of 1 gallon of water at that temperature is, by definition, 1 ml. Thus, the density of water, for example, varies with temperature, and its specific gravity at equal temperatures is always unity. The standard temperatures for a specific gravity in the petroleum industry in North America are 60/60°F (15.6/15.6°C).

The *density* and *specific gravity* of heavy oil (ASTM D287, ASTM D1298, ASTM D941, ASTM D1217, ASTM D1555) are two properties that have found wide use in the industry for preliminary assessment of the character of the oil. In particular, heavy oil with a high content of asphaltenes and resins (low kerosene) and poor mobility at ambient temperature and pressure, thereby requiring vastly different processing sequences, may have a specific gravity (density) of about 0.95.

Specific gravity is influenced by chemical composition, but quantitative correlation is difficult to establish. Nevertheless, it is generally recognized that increased amounts of aromatic compounds result in an increase in density, whereas an increase in saturated compounds results in a decrease in density. It is also possible to recognize certain preferred trends between the API gravity of crude oils and residua and one or more of the other physical parameters. For example, a correlation exists between the API gravity and sulfur content, Conradson carbon residue, and viscosity (Speight, 2000 and references cited therein). However, the derived relationships between the density of heavy oil and its fractional composition are valid only when applied to a certain type of heavy oil and may lose their significance when applied to heavy oil from different sources.

The values for density (and specific gravity) cover an extremely narrow range, considering the differences in heavy oil behavior. In an attempt to delineate a more meaningful relationship between the physical properties and processability of the various crude oils, the American Petroleum Institute devised a measurement of gravity based upon the Baumé scale for industrial liquids. The Baumé scale for liquids lighter than water was used initially:

$$^{\circ}\text{Baumé} = 140/\text{sp gr @ } 60/60^{\circ}\text{F} - 130$$

However, a considerable number of hydrometers calibrated according to the Baumé scale were found at an early period to be in error by a consistent amount, and this led to the adoption of the equation

$$^{\circ}\text{API} = 141.5/\text{sp gr @ } 60/60^{\circ}\text{F} - 131.5$$

The specific gravity of conventional crude oil usually ranges from about 0.8 (45.3° API) for conventional crude oil to about 1.0 (10° API) for heavy oil. This is in keeping with the general trend that a lower atomic hydrogen/carbon ratio (increased aromaticity) leads to a decrease in API gravity (or, more correctly, an increase in specific gravity).

Density, specific gravity, or API gravity may be measured, depending upon the properties of the heavy oil sample, by means of a hydrometer (ASTM D287, ASTM D1298) or by means of a pycnometer (ASTM D941, ASTM D1217). The variation of density with temperature, effectively the coefficient of expansion, is a property of great technical importance since most crude oils are sold by volume, and specific gravity is usually determined at the prevailing temperature (21°C, 70°F) rather than at the standard temperature (60°F, 15.6°C). The tables of gravity corrections (ASTM D1555) are based on an assumption that the coefficient of expansion is a function (at fixed temperatures) of density only.

4.1.5 Viscosity

Viscosity is the force in dynes required to move a plane of 1 cm^2 area at a distance of 1 cm from another plane of 1 cm^2 area through a distance of 1 cm in 1 second. In the centimeter-gram-second system, the unit of viscosity is the *poise* (P) or *centipoise* (cP) (1 cP = 0.01 P). Two other terms in common use are *kinematic viscosity* and *fluidity*. The kinematic viscosity is the viscosity in centipoise divided by the specific gravity. The unit of kinematic viscosity is the *stoke* (cm^2/s), although *centistokes* (0.01 cSt) is in more common usage. Fluidity is simply the reciprocal of viscosity.

The viscosity of heavy oils is a critical property in predicting oil recovery. In fact, viscosity is often cited as the single most important fluid characteristic governing the motion of crude oil. It is actually a measure of the internal resistance to motion of a fluid by reason of the forces of cohesion between molecules or molecular groupings. It is unfortunate that oil-rock interactions and reservoir structure are often omitted when the focus is on oil viscosity.

The viscosity (ASTM D445, ASTM D88, ASTM D2161, ASTM D341, ASTM D2270) of heavy oil varies markedly over a very wide range. Values vary from several hundred centipoises at room temperature to many thousands of centipoises at the same temperature.

Many types of instruments have been proposed for the determination of viscosity, but as in the determination of density, the choice of an instrument depends upon the properties of the oil.

The simplest and most widely used instruments are capillary types (ASTM D445), and the viscosity is derived from the equation

$$\mu = Br4P/8nl$$

where r is the tube radius, l the tube length, P the pressure difference between the ends of a capillary, n the *coefficient of viscosity*, and μ the quantity discharged in unit time. Not only are such capillary instruments the most simple, but when designed in accordance with known principle and used with known necessary correction factors, they are probably the most accurate viscometers available. It is usually more convenient, however, to use relative measurements. For this purpose, the instruments are calibrated with an appropriate standard liquid of known viscosity.

Batch flow times are generally used; in other words, the time required for a fixed amount of sample to flow from a reservoir through a capillary is the datum that is actually observed. Any features of technique that contribute to longer flow times are usually desirable. Some of the principal capillary viscometers in use are those of Cannon-Fenske, Ubbelohde, Fitzsimmons, and Zeitfuchs.

The *Saybolt universal viscosity* (SUS) (ASTM D88) is the time in seconds required for the flow of 60 ml of heavy oil, at constant temperature, through a calibrated orifice. The *Saybolt furol viscosity* (SFS) (ASTM D88) is determined in a similar manner except that a larger orifice is employed.

As a result of the various methods for viscosity determination, it is not surprising that much effort has been spent on interconversion of the several scales, especially converting Saybolt to kinematic viscosity (ASTM D2161),

$$\text{Kinematic viscosity} = \text{a} \times \text{Saybolt } s + \text{b/Saybolt } s$$

where a and b are constants.

The SUS equivalent to a given kinematic viscosity varies slightly with the temperature at which the determination is made because the temperature of the calibrated receiving flask used in the Saybolt method is not the same as that of the oil. Conversion factors are used to convert kinematic viscosity from 2 to 70 cSt at 38°C (100°F) and 99°C (210°F) to equivalent Saybolt universal viscosity in seconds (ASTM, 2008). Appropriate multipliers are listed to convert kinematic viscosity over 70 cSt. For a kinematic viscosity determined at any other temperature, the equivalent Saybolt universal value is calculated by use of the Saybolt equivalent at 38°C (100°F) and a multiplier that varies with the temperature:

$$\text{Saybolt } s \text{ at } 100°\text{F } (38°\text{C}) = \text{cSt} \times 4.635$$

$$\text{Saybolt } s \text{ at } 210°\text{F } (99°\text{C}) = \text{cSt} \times 4.667$$

Various studies have also been conducted on the effect of temperature on viscosity since the viscosity of heavy oil decreases as the temperature increases:

$$\log \log (n + c) = A + B \log T$$

where n is absolute viscosity, T is temperature, and A and B and c are constants. This equation has been sufficient for most purposes and has come into very general use. The constants A and B vary widely with different oils, but c remains fixed at 0.6 for all oils having a viscosity over 1.5 cSt (it increases only slightly at lower viscosity [0.75 at 0.5 cSt]). The viscosity-temperature characteristics of any oil, so plotted, thus create a straight line, and the parameters A and B are equivalent to the intercept and slope of the line. To express the viscosity and viscosity-temperature characteristics of an oil, the slope and the viscosity at one temperature must be known. The usual practice is to select 38°C (100°F) and 99°C (210°F) as the observation temperatures.

Suitable conversion tables are available (ASTM D341). Each table or chart is constructed in such a way that, for any given heavy oil, the viscosity-temperature points result in a straight line over the applicable temperature range. Thus, only two viscosity measurements need be made, at temperatures far enough apart, to determine a line on the appropriate chart from which the approximate viscosity at any other temperature can be read.

4.2 Thermal Properties

4.2.1 Carbon Residue

The carbon residue (ASTM D189 and ASTM D524) of heavy oil is a property that can be correlated with several other properties of the oil. It may be used to evaluate the carbonaceous depositing characteristics of heavy oil during thermal recovery.

There are two older, well-used methods for determining the carbon residue: the Conradson method (ASTM D189) and the Ramsbottom method (ASTM D524). Both are equally applicable to heavy oil, but the metallic constituents will give erroneously high carbon residues. The metallic constituents must first be removed from the oil or they can be estimated as ash by complete burning of the coke after carbon residue determination. There is no exact correlation between the two methods but it is possible to interconnect the data (ASTM, 2007).

Another method (ASTM D4530) requires smaller sample amounts and was originally developed as a *thermogravimetric method*. The carbon residue produced by this method is often referred to as the *micro-carbon residue* (*MCR*). Agreements between the data from the this and the Conradson and Ramsbottom methods are good, making it possible to interrelate all of the data from carbon residue tests (Long and Speight, 1989).

4.2.2 Specific Heat

Specific heat is the quantity of heat required to raise a unit mass of material through one degree of temperature (ASTM D2766). It is an extremely important engineering quantity in practice and is used in all calculations on heating and cooling heavy oil. Many measurements have been made on various hydrocarbon materials, but the data for most purposes may be summarized by the general equation

$$C = 1/d\,(0.388 + 0.00045t)$$

where C is the specific heat at t°F of an oil whose specific gravity 60/60°F is d. Thus, specific heat increases with temperature and decreases with specific gravity.

4.2.3 Heat of Combustion

The gross heat of combustion of heavy oil is given with a reasonable degree of accuracy by the equation

$$Q = 12{,}400 - 2100d^2$$

where d is the 60/60°F specific gravity. Deviation is generally less than 1%, although highly aromatic heavy oil may show considerably higher values.

For thermodynamic calculation of equilibria, combustion data of extreme accuracy are required because the heats of formation of water and carbon dioxide are large in comparison with those of the hydrocarbons. Great accuracy is also required of the specific heat data for the calculation of free energy or entropy. Much care must be exercised in selecting values from the literature for these purposes since many of those available were determined before the development of modern calorimetric techniques.

4.2.4 Volatility

The volatility of a liquid or liquefied gas may be defined as its tendency to vaporize, that is, to change from the liquid to the vapor or gaseous state. Because one of the three essentials for combustion in a flame is that the fuel be in the gaseous state, volatility is a primary characteristic of liquid fuels. The distillation profile is also a measure of the relative amounts of these liquid fuels (albeit small and unrefined) in heavy oil.

Similarly, there must be some estimate of the ability of the constituents of heavy oil to distill, or steam distill, from the oil during thermal methods of enhanced oil recovery. However, before any volatility tests are carried out, it must be recognized that the presence of more then 0.5% water in test samples of heavy oil can cause several problems during distillation procedures. Water has a high heat of vaporization, necessitating the application of additional thermal energy to the distillation flask. Water is relatively easily superheated, and therefore excessive *bumping* can occur, leading to erroneous readings and the real potential for destruction of the glass equipment. In addition, steam formed during distillation can act as a carrier gas, and high-boiling-point components may end up in the distillate (often referred to as *steam distillation*).

Centrifugation can be used to remove water (and sediment) if the sample is not a tight emulsion. Other methods that are used to remove water include:

1. Heating in a pressure vessel to control loss of light ends

2. Addition of calcium chloride as recommended in ASTM D1160

3. Addition of an azeotroping agent such as *iso*-propanol or *n*-butanol

4. Removal of water in a preliminary low-efficiency or flash distillation followed by reblending of the hydrocarbon which co-distills with the water into the sample (see also IP 74)

5. Separation of the water from the hydrocarbon distillate by freezing

The vaporizing tendencies of petroleum and petroleum products are the basis for the general characterization of liquid petroleum fuels (ASTM D2715). A test (ASTM D6) also exists for determining the loss of material when crude oil and heavy oil are heated. Another test method (ASTM D20), for the distillation of road tars, might also be applied to estimating the volatility of high-molecular-weight unknown residues.

For some purposes, it is necessary to have information on the initial stage of vaporization and the potential hazards, even with heavy oil, that such a property can cause. To supply this need, flash-and-fire, vapor pressure, and evaporation methods are available. The data from the early stages of the several distillation methods are also useful. For other uses, it is important to know the tendency of a product to partially vaporize or to completely vaporize, and in some cases, to know if small quantities of high-boiling components are present. For such purposes, chief reliance is placed on the distillation methods.

The *flash point* of petroleum or a petroleum product is the temperature to which the product must be heated under specified conditions to give off sufficient vapor to form a mixture with air that can be ignited momentarily by a specified flame (ASTM D56, ASTM D92, ASTM D93, ASTM D1310, ASTM D3828).

The Pensky-Marten apparatus using a closed or open system (ASTM D93) is the standard instrument for measuring flash points above 50°C. The Cleveland Open-Cup method (ASTM D92) is used for the determination of the *fire point* (the temperature at which the sample will ignite and burn for at least five seconds).

The Pensky-Marten apparatus consists of a brass cup mounted in an air bath and heated by a gas flame. A propeller-type stirrer, operated by a flexible drive, extends from the center of the cover into the cup. The cover has four openings, one for the insertion of a thermometer and the others, fitted with sliding shutters, for the introduction of a pilot flame and for ventilation. The temperature of the oil in the cup is raised at 5 to 6°C/minute (9 to 11°F/minute). The stirrer is rotated at approximately 60 rpm. When the temperature has risen to approximately 15°C (27°F) from the anticipated flash point, the pilot flame is dipped into the oil vapor for two seconds for every 1°C (1.8°F) rise in temperature up to 105°C (221°F). Above 105°C (221°F), the flame is introduced every 2°C (3.6°F) rise in temperature. The flash point is the temperature at which a distinct flash is observed when the pilot flame meets the vapor in the cup. The *open-cup flash point* is determined after the closed-cup flash point by removing the cover and continuing the heating until a distinct flash occurs across the open cup.

The Abel *closed-cup* apparatus (IP 170) consists of a brass cup sealed in a small water bath that is immersed in a second water bath. The cover of the brass cup is fitted in a manner similar to that in the Pensky-Marten apparatus. For crude oils and products with a flash point less than 30°C (<86°F), the outer bath is filled with water at 55°C (131°F) and is not heated further. The oil under test is then placed inside the cup. When the temperature reaches 19°C (66°F), the pilot flame is introduced every 0.5°C (1°F) until a flash is obtained. For oils with flash points in excess of 30°C (>86°F) and less than 50°C (<122°F), the inner water bath is filled with cold water to a depth of 35 mm. The outer bath is filled with cold water and heated at a rate of 1°C/minute (1.8°F/minute). The flash point is obtained as described before.

From the viewpoint of safety, information about the flash point is of most significance at or slightly above the maximum temperatures (30-60°C, 86-140°F) that may be encountered in storage, transportation, and use of crude oil, heavy oil, and their products, in either closed or open containers. In this temperature range, the relative fire and explosion hazard can be estimated from the flash point. For products with flash point below 40°C (104°F), special precautions are necessary for safe handling. Flash points above 60°C (140°F) gradually lose their safety significance until they become indirect measures of some other quality.

The flash point of heavy oil can also be used to detect contamination. A substantially lower flash point than expected for a product is a reliable indicator that a product has become contaminated with a more volatile product, such as gasoline. The flash point is also an aid in establishing the identity of a particular hydrocarbon contaminant.

A further aspect of volatility that receives considerable attention is the *vapor pressure* of heavy oil, which may be close to zero. The vapor pressure is the force exerted on the walls of a closed container by the vaporized portion of a liquid. Conversely, it is the force that must be exerted on the liquid to prevent it from vaporizing further (ASTM D323). The vapor pressure increases with temperature. The temperature at which the vapor pressure of a liquid, either a pure compound or a mixture of many compounds, equals one atmosphere pressure (14.7 psi, absolute) is designated as the boiling point of the liquid.

Heavy oil can be subdivided by distillation into a variety of fractions of different *boiling ranges* (*cut points*) using a variety of standard methods specifically designed for this task (the lower boiling fractions are de-emphasized because of the nature of heavy oil):

ASTM D86 (IP 123)—Distillation of Petroleum Products

ASTM D285—Distillation of Crude Petroleum

ASTM D1160—Distillation of Petroleum Products at Reduced Pressure

ASTM D2887—Test Method for Boiling Range Distribution of Petroleum Fractions by Gas Chromatography

ASTM D 2892—Distillation of Crude Petroleum (15 Theoretical Plate Column).

Distillation involves the general procedure of vaporizing the petroleum liquid in a suitable flask either at *atmospheric pressure* (ASTM D86, ASTM D285, ASTM D447, ASTM D2892) or at *reduced pressure* (ASTM D1160). There are also test methods for the distillation of pitch (ASTM D2569) and cutback asphalt (ASTM D402) that can be applied to heavy oil. However, most of the methods specify an upper atmospheric equivalent temperature (AET) limit of 360°C (680°F) and therefore are too limited to be of value in the analysis of low-volatility API gravity heavy oils.

In the simplest case, the distillation method involves using a standard round-bottom distillation flask of 250 ml capacity attached to a water-cooled condenser. The thermometer bulb is placed at the opening to the side arm of the flask. One hundred milliliters of sample are placed in the flask and heated by a small gas flame so as to produce 10 milliliters of distillate every four or five minutes. The temperature of initial distillation. the temperature at which each further 10 ml distils, and the final boiling point are all recorded.

The three standard distillation procedures that are commonly used by laboratories, either singly or in combination, to determine the distillation curve of heavy oil or to produce fractions from heavy oil are:

> ASTM D2892—Distillation of Crude Petroleum (15 Theoretical Plate Column)
>
> ASTM D1160—Distillation of Petroleum Products at Reduced Pressures
>
> ASTM D5236—Distillation of Heavy Hydrocarbon Mixtures (Vacuum Potstill Method)

The *ASTM D2892 method* describes the procedure for distilling crude petroleum up to 400°C AET. This method is often referred to as the *true-boiling-point (TBP) distillation* method and is adequate for oil when an estimation of volatiles and nonvolatiles is all that is required.

The *ASTM D1160 method* is used to determine the boiling ranges of petroleum products to a maximum liquid temperature of 400°C (752°F) at pressures as low as 1 mm Hg and is suitable for use with heavy oil. In this method, a 200 ml sample is weighed to the nearest 0.1 g in a distillation flask. The distillation assembly is evacuated to the desired pressure, and heat is applied to the flask as rapidly as possible using a 750-watt heater. When refluxing liquid appears, the rate of heating is adjusted so that the distillate is recovered at 4 to 8 ml/minute until the distillation is complete. However, because of the thermal sensitivity of heavy oil, cracking will most likely occur before the liquid temperature reaches 400°C (752°F). An increase in distillation rate accompanied by a drop in head temperature, loss of vacuum in the system that is restored when heat to the still is reduced, and production of vapor clouds in the system are all evidence of cracking.

The *vacuum potstill method* (ASTM D5236) is a procedure for the distillation of heavy oil samples having initial boiling points above 150°C (300°F). The method employs a pot still with a low-pressure drop entrainment separator. The method also provides for the determination of standard distillation profiles to the highest atmospheric equivalent temperature possible by conventional distillation.

In the *spinning band method,* fractions of feedstocks such as heavy oil and residua with initial boiling points above room temperature at atmospheric pressure can be prepared. For such materials, the initial boiling point of the sample should exceed room temperature at atmospheric pressure. The distillation is terminated at an atmospheric equivalent temperature of 524°C (975°F). For heavy oil samples, the distillation is terminated at a pot temperature of 360°C (680°F).

Samples are distilled under atmospheric and reduced pressures of 20.0 and 0.60 mm Hg (2.67 kPa and 80 Pa) in a distillation unit equipped with a spinning band column. The spinning band, which effectively provides a large contact area between the liquid phase and the vapor phase, increases the number of theoretical plates in the column and thus its fractionating efficiency. Readings of vapor temperature (which is convertible to atmospheric equivalent temperature) and distillate volume (which is convertible to per cent by volume) are used to plot a distillation curve. Distillate yields for naphtha, light gas oil, heavy gas oil, and residue fractions are determined on a gravimetric basis.

Short path distillation produces a single distillate and a single residue fraction defined by the operating temperature and pressure of the still. This procedure is used to generate high-boiling-point fractions with end points up to 700°C (1290°F) for further analysis. Since only one cut temperature is used in each run, generation of a distillation curve using this equipment would be time consuming. In this method, the material to be fractionated is introduced at a constant rate onto the hot inner wall of the evaporator under high vacuum. Rotating (teflon) rollers ensure that the film on the wall is kept thin. The feedstock is progressively distilled at the fixed conditions of temperature and pressure. The distillate vapors condense on a concentric cold surface (60°C, 140°F) placed a short distance from the hot wall inside the still. The condensate then drains by gravity to the base of the *cold finger* where it is collected. The residue drains down the hot wall and is collected through a separate port.

Short path distillation permits collection of deeper cut distillate and residue fractions than conventional vacuum distillation methods. Atmospheric equivalent temperature end points of up to 660°C (1220°F) can be attained without thermal cracking of the material. The distillation, however, is conducted under non-equilibrium conditions, so the cut point is not as specific as in other vacuum distillation methods.

Generally, the distillation tests are planned so that the data are reported in terms of one or more of the following items:

Initial boiling point is the thermometer reading in the neck of the distillation flask when the first drop of distillate leaves the tip of the condenser tube. This reading is materially affected by a number of test conditions, namely, room temperature, rate of heating, and condenser temperature.

Distillation temperatures are usually observed when the level of the distillate reaches each 10% mark on the graduated receiver, with the temperatures for the 5% and 95% marks often included. Conversely, the volume of the distillate in the receiver, that is, the percentage recovered, is often observed at specified thermometer readings.

End-point or *maximum temperature* is the highest thermometer reading observed during distillation. In most cases, it is reached when all of the sample has been vaporized. If a liquid residue remains in the flask after the maximum permissible adjustments are made in heating rate, this is recorded as indicative of the presence of very-high-boiling compounds.

Dry point is the thermometer reading at the instant the flask becomes dry. It is for special purposes, such as for solvents and for relatively pure hydrocarbons. For these purposes, dry point is considered more indicative of the final boiling point than end-point or maximum temperature.

Recovery is the total volume of distillate recovered in the graduated receiver and *residue* is the liquid material, mostly condensed vapors, left in the flask after it has been allowed to cool at the end of distillation. The residue is measured by transferring it to an appropriate small graduated cylinder. Low or abnormally high residues indicate the absence or presence, respectively, of high-boiling components.

Total recovery is the sum of the liquid recovery and residue; *distillation loss* is determined by subtracting the total recovery from 100%. Distillation loss is, of course, the measure of the portion of the vaporized sample that does not condense under the conditions of the test. Like the initial boiling point, distillation loss is affected materially by a number of test conditions, namely, condenser temperature, sampling and

> receiving temperatures, barometric pressure, heating rate in the early part of the distillation, and others. Provisions are made for correcting high distillation losses for the effect of low barometric pressure because of the practice of including distillation loss as one of the items in some fuel specifications.
>
> *Percentage evaporated* is the percentage of liquid recovered at a specific thermometer reading or other distillation temperatures, or the converse. The amounts that have evaporated are usually obtained by plotting observed thermometer readings against the corresponding observed recoveries plus, in each case, the distillation loss. The initial boiling point is plotted with the distillation loss as the percentage evaporated. Distillation data are reproducible, particularly for the more volatile products.

The actual procedure involves the distillation of the oil in one- to five-gallon quantities in a column equivalent to 14 theoretical plates, with reflux adjusted, to obtain a true boiling point versus yield relationship. Narrow fractions can be taken throughout the distillation, and the still pressure can be reduced as necessary to avoid decomposition in the higher boiling ranges. The fractionating distillation is continued to a relatively high temperature, such as 370°C (700°F) equivalent temperature at 760 mm of mercury absolute. The actual operating pressure of the still at this stage would be in the region of 2 mm of mercury absolute. The distillate fractions up to this point cover the light and middle distillates. After the analysis of the required fractions in this range, curves relating the various properties to yield on fraction or on crude oil can be drawn. A series of residues can also be prepared by appropriate blending of distillates with the final residue. After analysis, property relationships with yield can also be prepared for the residues.

From the true-boiling-point curve, the yields of products of a required boiling range can be deduced, and fractions can be blended to match required products for further analysis or treatments. The residue (b.p. >370°C, >700°F) obtained from the true-boiling-point distillation quoted contains the main lubricating oil stock. By further distillation under high vacuum, a series of distillates can be prepared to cover the various lubricating oil grades. With a waxy crude oil, it is necessary to remove the wax from these distillates and further examine the wax-free lubricating oil distillates.

It is often useful in this type of crude oil assay to be able to extend the boiling-point data to higher temperatures than are possible in the fractionating distillation method previously described. For this purpose, a vacuum distillation in a simple still, with no fractionating column (similar to the ASTM D1160), can be carried out. This distillation, which is done under fractionating conditions equivalent to one theoretical plate, allows the boiling-point data to be extended to about 600°C (1110°F) (corrected to 760 mm of mercury absolute) with many crude oils. Bearing in mind the limitations in accuracy of the temperature measurements in the high boiling ranges of crude oil, it is possible to construct a wide-range boiling-point curve for the crude oil by combining the results of the simple distillation with those of the main distillation. This method gives useful comparative and reproducible results that are often accurate enough for engineering purposes.

It is advantageous to any analytical, research, or development organization dealing with crude oil to collect information by the above methods on a standard series of distillates or residues from crude oils so that valuable comparisons become available as the data accumulate. These distillates should be prepared as near as possible to the boiling ranges that are known to be suitable for marketable products. The residue (>370°C, >700°F) can be further fractionated, if required, using a high-vacuum still of sufficient capacity to give fractions in suitable quantity for subsequent analysis.

Another method that is increasing in popularity for application to a variety of feedstocks is commonly known as *simulated distillation* (ASTM D2887). The term *simulated distillation by gas chromatography* (or *simdis*) is used throughout the industry to refer to this technique. This is a low-resolution, temperature-programmed gas chromatographic analysis intended to simulate the time-consuming true-boiling point distillation method. It is based on the observation that hydrocarbons generally are eluted from a nonpolar column in the order of their boiling point, with the retention time being converted to equivalent distillation temperatures. The accumulated detector response is related to the amount of sample distilled. In practice, the nonpolar stationary phases do discriminate to some extent among different classes of hydrocarbons. The retention time is a function of the adsorptive property of the stationary phase, the vapor pressure, and the heat of vaporization. The empirical correlation between the retention time and boiling point, usually established by running a series of n-alkanes, is not strictly valid for other hydrocarbon classes such as aromatic and cycloalkane derivatives, which show a tendency to elute earlier than n-alkanes having the same boiling points.

The simulated distillation method has been well researched in terms of method development and application (Hickerson, 1975; Green, 1976; Stuckey, 1978; Vercier and Mouton, 1979; Thomas et al., 1983; Romanowski and Thomas, 1985; MacAllister and DeRuiter, 1985; Schwartz et al., 1987; Thomas et al., 1987). The benefits of the technique include good comparisons with other ASTM distillation data as well as the application to higher-boiling fractions. In fact, data output includes the provision of the corresponding Engler profile (ASTM D86) as well as the prediction of other properties such as vapor pressure and flash point.

4.2.5 Liquefaction and Solidification

The liquefaction and solidification of heavy oil seems to draw little attention in the standard petroleum science textbooks. Yet these properties are very important in the handling of heavy oil, both at the wellhead and in the refinery. In fact, since heavy oil can be a borderline liquid or near solid at ambient temperature, problems may arise from solidification during normal use or storage.

The *melting point* test is a test (ASTM D87, ASTM D127) that is widely used by suppliers of wax and wax consumers and can also be applied to heavy oil. However, it is the test for the *softening point* (ASTM D36, ASTM D2398), defined as the temperature at which a disk of the material softens and sags downward a distance of 25 mm under the weight of a steel ball under strictly specified conditions, that finds wider use for heavy oil. The *dropping point* test (ASTM D566), a test nearly equivalent to the softening point test, is used for lubricating greases.

Recently, more emphasis has been placed on the *pour point,* which, in conjunction with the reservoir temperature, can give indications of the fluidity and mobility of the heavy oil in the reservoir. The pour point of a crude oil was originally applied to crude oil that had a high wax content. More recently, the pour point, like the viscosity, is determined principally for use in pumping arid pipeline design calculations.

To determine the pour point (ASTM D97), the sample is first heated to 46°C (115°F) and cooled in air to 32°C (90°F). The tube is then immersed in the same series of coolants as used for the determination of the *cloud point*. The sample is inspected at temperature intervals of 2°C (3°F) by withdrawal and holding horizontal for five seconds until no flow is observed during this time interval.

More recently, the pour point has also found use as an indicator of the temperature at which heavy oil will flow during in situ recovery operations (Figure 4–1). For example, for asphaltic crude oils where paraffin precipitation will not occur, if 21°C (70°F) is the pour point of heavy oil in a reservoir where the temperature is 38°C (100°F), the oil is liquid under reservoir conditions and will mobile (will flow) under those conditions. On the other hand, tar sand bitumen (pour point: 60°C, 140°F) in a deposit (temperature: 10°C, 50°F) will be solid and immobile. This state of the oil in the reservoir can also have consequences on the ability of gases and liquids (e.g., steam, hot water) used for recovery operations to penetrate the reservoir or deposit. Although pressure can have some influence on the pour point, the effect is not large and is unlikely to result in any general conclusions. Indeed, there is a relationship between API gravity and pour point. Any increase in pour point due to an increase in pressure (surface compared to reservoir or deposit pressure) will most likely be negated as the API gravity decreases with increase in temperature (60°F compared to reservoir temperature).

Whether or not the relationship of pour point and reservoir temperature is generally accepted remains to be seen, but the withdrawal from the use of a single parameter to predict oil behavior is a necessity. The use of two parameters, whatever they may be, gives a more realistic view of oil behavior.

4.2.6 Solubility

Although not truly a thermal property, the solubility parameter of crude oil and its constituent fractions is of interest during thermal methods of recovery. The solubility parameter of the crude oil fraction, especially the asphaltene fractions, has been the subject of study, and some interesting results have emerged (Speight, 2007). In fact, phase separation as can occur during thermal recovery of heavy oil can be explained by use of the solubility parameter, δ, for petroleum fractions and for the solvents. As an extension of this concept, there is sufficient data to draw a correlation between the atomic hydrogen/carbon ratio and the solubility parameter for hydrocarbons and the constituents of the lower-boiling fractions of petroleum. Hydrocarbon liquids can dissolve polynuclear hydrocarbons in which

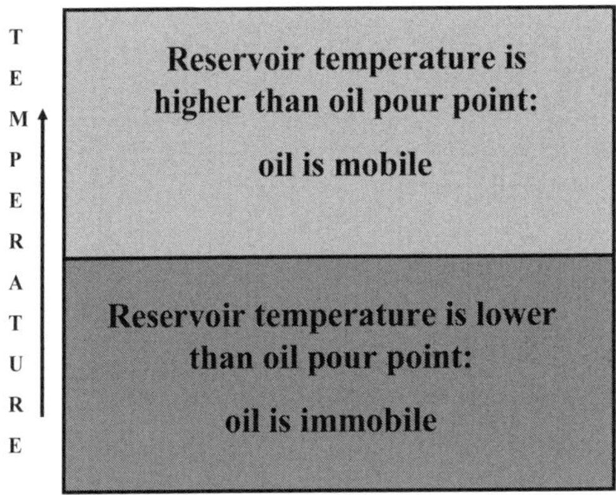

Figure 4–1 *Relationship of pour point and reservoir temperature.*

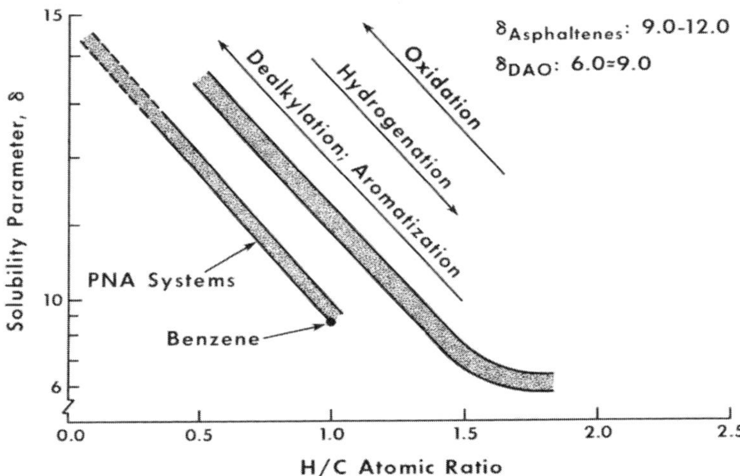

Figure 4–2 *Representation of the variation of the solubility parameter of petroleum fractions through variation with the H/C atomic ratio and comparison to benzene and polynuclear aromatic systems.*

there is usually less than a three-point difference between the lower solubility parameter of the solvent and the higher solubility parameter of the solute. Thus, a parallel, or near-parallel, line can be assumed that allows the solubility parameter of the asphaltenes and resins to be estimated (Figure 4–2).

By this means, the solubility parameter of asphaltenes can be estimated to fall in the range 9 to 12, which is in keeping with the asphaltenes being composed of a mixture of different compound types with a variation in polarity. As the thermal reaction proceeds (especially if the recovery process employs super-heated steam of combustion), it causes the removal of alkyl side chains from the asphaltenes, decreases the hydrogen-to-carbon atomic ratio (Figure 4–3), and increases the solubility parameter. Concurrently, changes occur to the oil medium, but they are of lesser overall effect and bringing about a higher solubility parameter differential between the reacted asphaltene constituents and resin constituents and the oil. As a result, deposition ensues.

The deposited reacted material is usually a product of the action of the highest-molecular-weight and/or the highest-polarity constituents in the asphaltene and resin fractions (Figure 4–4). This is of benefit to the refiner of the produced oil but may be a disadvantage insofar as the deposit causes blockage of the reservoir flow channels.

Another aspect of this reaction is the order of deposition relative to models applied to the system. The more polar constituents (e.g., the amphoteric constituents, Figure 4–4) of the asphaltene and resin fractions are more thermally labile than the lower-polarity constituents (e.g., the neutral polar constituents, Figure 4–4). As a result, products from the amphoteric constituents will exceed the solubility parameter differential more quickly and will separate from the oil medium first and at an earlier time than could be predicted if an average property is used for any model applied to the system.

4.3 Metals Content

Heteroatoms (*nitrogen, oxygen, sulfur,* and *metals*) are found in every crude oil; their concentrations have to be reduced to convert the oil to transportation fuel. If nitrogen and sulfur are present in the final fuel during combustion, nitrogen oxides (NOx) and sulfur oxides (SOx) form, respectively. In addition, metals affect many upgrading processes adversely, poisoning catalysts in refining and causing deposits in combustion.

Heteroatoms affect every aspect of refining. Sulfur is usually the most concentrated and is fairly easy to remove; many commercial catalysts are available that routinely remove 90% of the sulfur. Nitrogen is

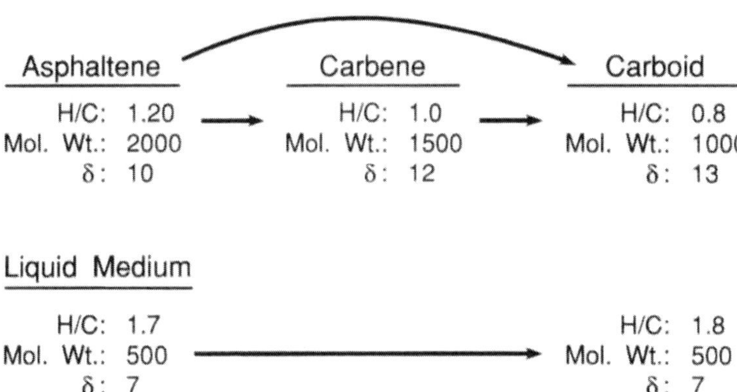

Figure 4–3 *Variation of the solubility parameter of the asphaltene fraction and the oil with reaction progress.*

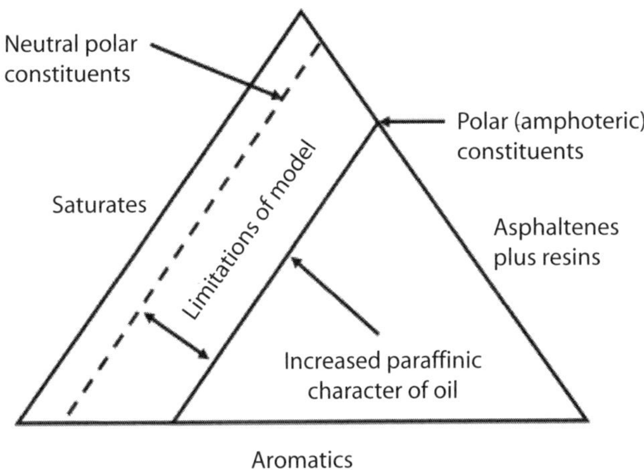

Figure 4–4 *Order of deposition of asphaltene constituents during thermal changes.*

more difficult to remove than sulfur, and there are fewer catalysts that are specific for nitrogen. Metals cause particular problems because they poison catalysts used for sulfur and nitrogen removal as well as other processes such as catalytic cracking. In addition, the levels of nickel and vanadium in heavy oil need to be evaluated by the geochemist when the origins of heavy oil are considered.

In other words, the trace components that are present in crude oils can produce adverse effects in refining; they can cause corrosion, affect the quality of refined products, or exert a deleterious influence on the efficiency of various processing catalysts. This last effect has become of increasing importance due to modern developments in refinery processing.

Heavy oils contain relatively high proportions of *metals* either in the form of salts or organometallic constituents (such as the metallo-porphyrins), which are extremely difficult to remove from the feed-stock. Indeed, the nature of the process by which residua are produced virtually dictates that all the metals in the original crude oil are concentrated in the residuum (Speight, 2000). Those metallic constituents that may actually *volatilize* under the distillation conditions and appear in the higher-boiling distillates are the exceptions. The deleterious effect of metallic constituents on the catalyst is known particularly through their ability to modify the selectivity of Zeolite catalysts, thereby causing an increase in the formation of coke at the expense of the more desirable liquid products. Thus, serious attempts have been made to develop catalysts that can tolerate a high concentration of metals without serious loss of catalyst activity or catalyst life. However, for the most part, the metals concentrate in the coke formed during thermal processes.

A number of the heavy metals such as nickel, vanadium, copper, and iron can also be effectively bound in large organic molecules characteristic of those found in the asphaltene fraction (the pentane- or heptane-insoluble portion of the feedstock) and resins. Nickel and vanadium porphyrins are commonly found and show high thermal stability, which allows them to pass through the extraction process into the upgrading process. Porphyrins are the major, but certainly not the only, organo-metallic complexes present. Metals may simply be entrapped or loosely bound in the very large molecules present in the asphaltenes and resins. Although *iron* is present as an organometallic compound, it occurs mostly in the form of process-accumulated rust or is scrounged from pipelines by the crude oil during shipping and pipelining. Thus, one should not worry about the geochemical significance of metal such as iron in crude oil.

Without doubt, these heavy metals are present in recovered heavy oil. Since catalysts are used extensively in upgrading and are readily poisoned by such metals, it is important to know the amounts present.

Significant levels typically found in heavy oil include 20 to 200 ppm of nickel and 50 to 500 ppm of vanadium.

Vanadium compounds can cause refractory damage in furnaces, adverse effects in glass manufacture, steel failure in turbines, as well as catalyst poisoning when present in distillate feedstocks. Arsenic and lead are also active catalyst poisons in reforming processes, and the presence of sodium in fuel oils causes failures in furnace brick-work. It is necessary, therefore, to examine crude oils and distillation unit feedstocks for the presence of these harmful contaminants.and to devise some form of treatment for reducing their effect during or before processing.

It is important to monitor process streams for metals content for several purposes:

- To track the degree of potential catalyst poisoning that may occur within the reactor due to feed metals content;
- To monitor the degree of catalyst physical or chemical breakdown into the product streams
- To provide one of many indicators of change in the process operation

The analysis of process stream solids can be accomplished by several instrumental techniques: inductively coupled argon plasma (ICAP) spectrometry, atomic absorption (AA) spectrometry, and X-ray fluorescence (XRF) spectrometry. Each technique has limitations in terms of sample preparation, sensitivity, sampling, time for analysis, and overall ease of use.

A variety of tests (ASTM D482, ASTM D1026, ASTM D1262, ASTM D1318, ASTM D1368, ASTM D1548, ASTM D1549, ASTM D2547, ASTM D2599, ASTM D2788, ASTM D3340, ASTM D3341, ASTM D3605), either directly or as the constituents of combustion ash, have been designated to determine the presence of metals in oil based on a variety of standard test methods. At the time of writing, a specific test for the determination of metals in whole feeds has not been designated, but this task can be accomplished by combustion of the sample so that only inorganic ash remains (ASTM D482). The ash can then be digested with an acid, and the solution can be examined for metal species by AA spectroscopy or ICAP spectrometry (ASTM C-1109, ASTM C-1111).

AA spectrometry provides very high sensitivity but requires careful sub-sampling, extensive sample preparation, and detailed sample-matrix corrections. XRF spectrometry requires little in terms of sample preparation but suffers from low sensitivity and major matrix corrections. ICAP spectrometry provides high sensitivity and few matrix corrections but requires a considerable amount of sample preparation depending on the process stream to be analyzed.

Whenever a technique requires ashing of the sample, one of the issues that arises is the potential loss of volatile nickel and vanadium compounds. Both sulfuric acid and elemental sulfur have been employed to prevent loss of these metals. In fact, because of this, a wet ashing method is often preferred over dry ashing (Wallace, 1988).

The analytical method should be selected depending on the sensitivity required, the compatibility of the sample matrix with the specific analysis technique, and the availability of facilities. Sample preparation, if it is required, can present problems. Significant losses can occur, especially in the case of organometallic complexes, and contamination of the environmental sample is of serious concern.

Analysis for metals in an organic matrix is most easily accomplished using XRF or neutron-activation techniques for quantitative analyses for metals. These particular techniques are also applicable for the direct analyses of oil sand solids. The direct aspiration of metal-containing hydrocarbons into the plasma instrument also provides quantitative information, with the assumption that any particulate present is included in the analyses. The precision of the analysis depends on the metal itself, the method used, and the standard used for calibration of the instrument. Relative standard deviations range from 1% to more than 20%.

In the inductively coupled argon plasma (ICAP) emission spectrometer method, nickel, iron, and vanadium content of gas oil samples in the range 0.1 mg/kg to 100 mg/kg. A 10 g sample of gas oil is charred with sulfuric acid and subsequently combusted to leave the ash residue. The resulting sulfates are then converted to their corresponding chloride salts to ensure complete solubility. A barium internal standard is added to the sample before analysis.

The ICAP spectrometer is an instrument routinely used for the analysis of the dissolved metal composition of aqueous samples. Solid samples are prepared in such a way (ashing then fusion) that their final analysis matrix is in the aqueous dissolved form. The analysis of metals in a hydrocarbon matrix, potentially containing solids important to the analysis, is a problem. The ashing of a hydrocarbon sample containing metals (whether in organic complexes or inorganic complexes) can lead to analysis errors; some of the complexes, specifically the organometallic compounds, are in a volatile form at typical ashing temperatures (700-800°C, 1290-1470°F). The organic metal complexes tend to originate in the high-molecular-weight hydrocarbons found naturally in the feed or generated in the reactor. The inorganic metal complexes originate naturally in the feed, from the breakdown of catalyst, or from the formation of inorganic complexes under the reactor conditions. Without ashing and subsequent fusion, any solids present may not reach the analytical region of the instrument, resulting in analytical errors.

Using the ICAP method for the analysis of nickel, vanadium, and iron present counteracts the two basic issues arising from metals analysis. The most serious issue is that these metals are partly or totally in the form of volatile chemically stable porphyrin complexes, and extreme conditions are needed to destroy the complexes without losing the metal through volatilization of the complex. The second issue is that the alternate direct aspiration of the sample introduces large quantities of carbon into the plasma. This carbon causes marked and somewhat variable background changes in all direct measurement techniques.

The combined utilization of X-ray tomography and magnetic resonance techniques for quantification of heavy oil fluid properties has also been proposed (Goodarzi et al., 2005; Goodarzi et al., 2007) and offers another aspect of oil testing. The incremental benefit of the proposed nucleonic techniques is that they are claimed to provide more detailed information about heavy oil, compared to conventional PVT measurements, which improves understanding of the nature of the foamy oil response.

4.4 References

ASTM. 2007. Annual Book of Standards. American Society for Testing and Materials, West Conshohocken, Pennsylvania.

ASTM C1109. Standard Test Method for Analysis of Aqueous Leachates from Nuclear Waste Materials Using Inductively Coupled Plasma-Atomic Emission Spectrometry. Annual Book of Standards. American Society for Testing and Materials, West Conshohocken, Pennsylvania.

ASTM C1111. Standard Test Method for Determining Elements in Waste Streams by Inductively Coupled Plasma-Atomic Emission Spectroscopy. Annual Book of Standards. American Society for Testing and Materials, West Conshohocken, Pennsylvania.

ASTM D6. Standard Test Method for Loss on Heating of Oil and Asphaltic Compounds. Annual Book of Standards. American Society for Testing and Materials, West Conshohocken, Pennsylvania.

ASTM D20. Standard Test Method for Distillation of Road Tars. Annual Book of Standards. American Society for Testing and Materials, West Conshohocken, Pennsylvania.

ASTM D36. Standard Test Method for Softening Point of Bitumen (Ring-and-Ball Apparatus). Annual Book of Standards. American Society for Testing and Materials, West Conshohocken, Pennsylvania.

ASTM D56. Standard Test Method for Flash Point by Tag Closed Cup Tester. Annual Book of Standards. American Society for Testing and Materials, West Conshohocken, Pennsylvania.

ASTM D86. Standard Test Method for Distillation of Petroleum Products at Atmospheric Pressure. Annual Book of Standards. American Society for Testing and Materials, West Conshohocken, Pennsylvania.

ASTM D87. Standard Test Method for Melting Point of Petroleum Wax (Cooling Curve). Annual Book of Standards. American Society for Testing and Materials, West Conshohocken, Pennsylvania.

ASTM D88. Standard Test Method for Saybolt Viscosity. Annual Book of Standards. American Society for Testing and Materials, West Conshohocken, Pennsylvania.

ASTM D92. Standard Test Method for Flash and Fire Points by Cleveland Open Cup Tester. Annual Book of Standards. American Society for Testing and Materials, West Conshohocken, Pennsylvania.

ASTM D93. Standard Test Methods for Flash Point by Pensky-Martens Closed Cup Tester. Annual Book of Standards. American Society for Testing and Materials, West Conshohocken, Pennsylvania.

ASTM D97. Standard Test Method for Pour Point of Petroleum Products. Annual Book of Standards. American Society for Testing and Materials, West Conshohocken, Pennsylvania.

ASTM D127. Standard Test Method for Drop Melting Point of Petroleum Wax, Including Petrolatum. Annual Book of Standards. American Society for Testing and Materials, West Conshohocken, Pennsylvania.

ASTM D189. Standard Test Method for Conradson Carbon Residue of Petroleum Products. Annual Book of Standards. American Society for Testing and Materials, West Conshohocken, Pennsylvania.

ASTM D285. Standard Test Method for Distillation of Crude Petroleum (Withdrawn 1985). Annual Book of Standards. American Society for Testing and Materials, West Conshohocken, Pennsylvania.

ASTM D287. Standard Test Method for API Gravity of Crude Petroleum and Petroleum Products (Hydrometer Method). Annual Book of Standards. American Society for Testing and Materials, West Conshohocken, Pennsylvania.

ASTM D323. Standard Test Method for Vapor Pressure of Petroleum Products (Reid Method). Annual Book of Standards. American Society for Testing and Materials, West Conshohocken, Pennsylvania.

ASTM D341. Standard Test Method for Viscosity-Temperature Charts for Liquid Petroleum Products. Annual Book of Standards. American Society for Testing and Materials, West Conshohocken, Pennsylvania.

ASTM D402. Standard Test Method for Distillation of Cut-Back Asphaltic (Bituminous) Products. Annual Book of Standards. American Society for Testing and Materials, West Conshohocken, Pennsylvania.

ASTM D445. Standard Test Method for Kinematic Viscosity of Transparent and Opaque Liquids (and the Calculation of Dynamic Viscosity). Annual Book of Standards. American Society for Testing and Materials, West Conshohocken, Pennsylvania.

ASTM D447. Test Method for Distillation of Plant Spray Oils (Withdrawn 1997). Annual Book of Standards. American Society for Testing and Materials, West Conshohocken, Pennsylvania.

ASTM D482. Standard Test Method for Ash from Petroleum Products. Annual Book of Standards. American Society for Testing and Materials, West Conshohocken, Pennsylvania.

ASTM D524. Standard Test Method for Ramsbottom Carbon Residue of Petroleum Products. Annual Book of Standards. American Society for Testing and Materials, West Conshohocken, Pennsylvania.

ASTM D566. Standard Test Method for Dropping Point of Lubricating Grease. Annual Book of Standards. American Society for Testing and Materials, West Conshohocken, Pennsylvania.

ASTM D941. Standard Test Method for Density and Relative Density (Specific Gravity) of Liquids by Lipkin Bicapillary Pycnometer (Withdrawn 1993). Annual Book of Standards. American Society for Testing and Materials, West Conshohocken, Pennsylvania.

ASTM D954. Standard Test Methods of Test for Apparent Density and Bulk Factor of Nonpouring Molding Powers (Withdrawn 1961). Annual Book of Standards. American Society for Testing and Materials, West Conshohocken, Pennsylvania.

ASTM D1026. Standard Test Method of Test for Sodium in Lubricating Oils and Additives (Gravimetric Method) (Withdrawn 1990). Annual Book of Standards. American Society for Testing and Materials, West Conshohocken, Pennsylvania.

ASTM D1160. Standard Test Method for Distillation of Petroleum Products at Reduced Pressure. Annual Book of Standards. American Society for Testing and Materials, West Conshohocken, Pennsylvania.

ASTM D1217. Standard Test Method for Density and Relative Density (Specific Gravity) of Liquids by Bingham Pycnometer. Annual Book of Standards. American Society for Testing and Materials, West Conshohocken, Pennsylvania.

ASTM D1262. Method of Test for Lead in New and Used Greases (Withdrawn 1990). Annual Book of Standards. American Society for Testing and Materials, West Conshohocken, Pennsylvania.

ASTM D1298. Standard Test Method for Density, Relative Density (Specific Gravity), or API Gravity of Crude Petroleum and Liquid Petroleum Products by Hydrometer Method. Annual Book of Standards. American Society for Testing and Materials, West Conshohocken, Pennsylvania.

ASTM D1310. Standard Test Method for Flash Point and Fire Point of Liquids by Tag Open-Cup Apparatus. Annual Book of Standards. American Society for Testing and Materials, West Conshohocken, Pennsylvania.

ASTM D1318. Standard Test Method for Sodium in Residual Fuel Oil (Flame Photometric Method). Annual Book of Standards. American Society for Testing and Materials, West Conshohocken, Pennsylvania.

ASTM D1368. Test Method for Trace Concentrations of Lead in Primary Reference Fuels (Withdrawn 1994). Annual Book of Standards. American Society for Testing and Materials, West Conshohocken, Pennsylvania.

ASTM D1548. Standard Test Method for Vanadium in Navy Special Fuel Oil (Withdrawn 1997). Annual Book of Standards. American Society for Testing and Materials, West Conshohocken, Pennsylvania.

ASTM D1549. Standard Test Method of Test for Zinc in Lubricating Oils and Additives (Polarographic Method) (Withdrawn 1984). Annual Book of Standards. American Society for Testing and Materials, West Conshohocken, Pennsylvania.

ASTM D1555. Standard Test Method for Calculation of Volume and Weight of Industrial Aromatic Hydrocarbons and Cyclohexane. Annual Book of Standards. American Society for Testing and Materials, West Conshohocken, Pennsylvania.

ASTM D1796. Standard Test Method for Water and Sediment in Fuel Oils by the Centrifuge Method (Laboratory Procedure). Annual Book of

Standards. American Society for Testing and Materials, West Conshohocken, Pennsylvania.

ASTM D2008. Method of Test for Characteristic Groups in Rubber Extender and Processing Oils by the Precipitation Method (Withdrawn 1975). Annual Book of Standards. American Society for Testing and Materials, West Conshohocken, Pennsylvania.

ASTM D2161. Standard Practice for Conversion of Kinematic Viscosity to Saybolt Universal Viscosity or to Saybolt Furol Viscosity. Annual Book of Standards. American Society for Testing and Materials, West Conshohocken, Pennsylvania.

ASTM D2270. Standard Practice for Calculating Viscosity Index From Kinematic Viscosity at 40 and 100°C. Annual Book of Standards. American Society for Testing and Materials, West Conshohocken, Pennsylvania.

ASTM D2398. Test Method for Softening Point of Bitumen in Ethylene Glycol (Ring-and-Ball) (Withdrawn 1984). Annual Book of Standards. American Society for Testing and Materials, West Conshohocken, Pennsylvania.

ASTM D2547. Method of Test for Lead in Gasoline, Columetric Chromate Method (Withdrawn 1989). Annual Book of Standards. American Society for Testing and Materials, West Conshohocken, Pennsylvania.

ASTM D2569. Standard Test Method for Distillation of Pitch. Annual Book of Standards. American Society for Testing and Materials, West Conshohocken, Pennsylvania.

ASTM D2599. Methods of Test for Lead in Gasoline by X-Ray Spectrometry (Withdrawn 1992). Annual Book of Standards. American Society for Testing and Materials, West Conshohocken, Pennsylvania.

ASTM D2662. Standard Specification for Polybutylene (PB) Plastic Pipe (SIDR-PR) Based on Controlled Inside Diameter (Withdrawn 2003). Annual Book of Standards. American Society for Testing and Materials, West Conshohocken, Pennsylvania.

ASTM D2715. Standard Test Method for Volatilization Rates of Lubricants in Vacuum. Annual Book of Standards. American Society for Testing and Materials, West Conshohocken, Pennsylvania.

ASTM D2766. Standard Test Method for Specific Heat of Liquids and Solids. Annual Book of Standards. American Society for Testing and Materials, West Conshohocken, Pennsylvania.

ASTM D2788. Method of Test for Trace Metals in Gas Turbine Fuels (Atomic Absorption Method) (Withdrawn 1983). Annual Book of Standards. American Society for Testing and Materials, West Conshohocken, Pennsylvania.

ASTM D2887. Standard Test Method for Boiling Range Distribution of Petroleum Fractions by Gas Chromatography. Annual Book of Standards. American Society for Testing and Materials, West Conshohocken, Pennsylvania.

ASTM D2892. Standard Test Method for Distillation of Crude Petroleum (15-Theoretical Plate Column). Annual Book of Standards. American Society for Testing and Materials, West Conshohocken, Pennsylvania.

ASTM D3178. Standard Test Methods for Carbon and Hydrogen in the Analysis Sample of Coal and Coke (Withdrawn 2007). Annual Book of Standards. American Society for Testing and Materials, West Conshohocken, Pennsylvania.

ASTM D3340. Standard Test Method for Lithium and Sodium in Lubricating Greases by Flame Photometer. Annual Book of Standards. American Society for Testing and Materials, West Conshohocken, Pennsylvania.

ASTM D3341. Standard Test Method for Lead in Gasoline-Iodine Monochloride Method. Annual Book of Standards. American Society for Testing and Materials, West Conshohocken, Pennsylvania.

ASTM D3605. Standard Test Method for Trace Metals in Gas Turbine Fuels by Atomic Absorption and Flame Emission Spectroscopy. Annual Book of Standards. American Society for Testing and Materials, West Conshohocken, Pennsylvania.

ASTM D3828. Standard Test Methods for Flash Point by Small Scale Closed Cup Tester. Annual Book of Standards. American Society for Testing and Materials, West Conshohocken, Pennsylvania.

ASTM D4006. Standard Test Method for Water in Crude Oil by Distillation. Annual Book of Standards. American Society for Testing and Materials, West Conshohocken, Pennsylvania.

ASTM D4007. Standard Test Method for Water and Sediment in Crude Oil by the Centrifuge Method (Laboratory Procedure). Annual Book of Standards. American Society for Testing and Materials, West Conshohocken, Pennsylvania.

ASTM D4530. Standard Test Method for Determination of Carbon Residue (Micro Method). Annual Book of Standards. American Society for Testing and Materials, West Conshohocken, Pennsylvania.

ASTM D5236. Standard Test Method for Distillation of Heavy Hydrocarbon Mixtures (Vacuum Potstill Method). Annual Book of Standards. American Society for Testing and Materials, West Conshohocken, Pennsylvania.

ASTM E777. Standard Test Method for Carbon and Hydrogen in the Analysis Sample of Refuse-Derived Fuel. Annual Book of Standards. American Society for Testing and Materials, West Conshohocken, Pennsylvania.

Goodarzi, N., Bryan, J., Mai, A., and Kantzas, A. 2005. Proceedings. SPE/PS-CIM/CHOA International Thermal Operations and Heavy Oil Symposium. Calgary, Alberta, Canada, November 1–3.

Goodarzi, N., Bryan, J., Mai, A., and Kantzas, A. 2007. SPE Journal. 12:305–315.

Green, L.E. 1976. Hydrocarbon Processing. 55(5):205.

Hickerson, J.F. 1975. In Special Publication No. STP 577. American Society for Testing and Materials, Philadelphia, p. 71.

IP 74. Water Content of Petroleum Products. Standard Test Method. Energy Institute (formerly Institute of Petroleum), London, England.

IP 170. Flash Point—Abel Closed Cup Method. Standard Test Method. Energy Institute (formerly Institute of Petroleum), London, England.

Long, R.B., and Speight, J.G. 1989. Studies in Petroleum Composition. I

Development of a Compositional Map for Various Feedstocks. Revue de l'Institut Français du Petrole. 44:205.

MacAllister, D.J., and DeRuiter, R.A. 1985. Paper SPE 14335. 60th Annual Technical Conference, Society of Petroleum Engineers. Las Vegas, September 22–25.

Reynolds, J.G. 1998. In Petroleum Chemistry and Refining. J.G. Speight (Editor). Taylor & Francis, Washington, DC.

Romanowski, L.J., and Thomas, K.P. 1985. Report No. DOE/FE/60177-2326. United States Department of Energy, Washington, DC.

Schwartz, H.E., Brownlee, R.G., Boduszynski, M.M., and Su, F. 1987. Analytical Chemistry 59:1393.

Speight, J.G. 2000. The Desulfurization of Heavy Oils and Residua. 2nd Edition. Marcel Dekker Inc., New York.

Speight, J.G. 2001. Handbook of Petroleum Analysis. John Wiley & Sons Inc., New York.

Speight, J.G. 2007. The Chemistry and Technology of Petroleum. 4th Edition. CRC-Taylor and Francis Group, Boca Raton, Florida.

Stuckey, C.L. 1978. J. Chromatographic Science. 16:482.

Thomas, K.P., Barbour, R.V., Branthaver, J.F., and Dorrence, S.M. 1983. Fuel. 62:438.

Thomas, K.P., Harnsberger, P.M., and Guffey, F.D. 1987. Report No. DOE/MC/11076-2451. United States Department of Energy. Washington, DC.

Vercier, P., and Mouton, M. 1979. Oil and Gas Journal. 77(38):121.

Wallace, E.D., ed. 1988. A Review of Analytical Methods for Bitumens and Heavy Oils. AOSTRA Technical Publication Series No. 5. Alberta Oil Sands Technology and Research Authority, Edmonton, Alberta, Canada.

CHAPTER 5

EXPLORATION AND GENERAL METHODS FOR OIL RECOVERY

Petroleum is found in the microscopic pores of sedimentary rocks such as sandstone and limestone. Not all of the pores in a rock will contain petroleum; some will be filled with water or brine that is saturated with minerals. Seismic surveys are used to try to predict where oil fields may be, but the only way of making certain is by drilling.

In their search for new recoverable oil and gas fields, petroleum companies use their vast knowledge of how and when petroleum was formed, their knowledge of geological structures that may have entrapped petroleum, and (of course) their knowledge of how to recover the resources.

Production rates from reservoirs depend on a number of factors, such as reservoir geometry (primarily formation thickness and reservoir continuity), pressure, and depth; rock type and permeability; fluid saturations and properties; extent of fracturing; number of wells and their locations; and the ratio of the permeability of the formation to the viscosity of the oil (Taber and Martin, 1983; DOE, 1996; Jayasekera, and Goodyear, 1999). Operators can increase production over that which would naturally occur by such methods as fracturing the reservoir to open new channels for flow, injecting gas and water to increase the reservoir pressure, or lowering oil viscosity with heat or chemicals. These supplementary techniques are expensive, and the extent to which they are used depends on such external factors as the operator's economic condition, sales prospects, and perceptions of future prices.

The extraordinary geological variability of different reservoirs means that production profiles differ from field to field. The natural drive force is a determining factor in production and, in some fields, an oil reservoir with the seemingly large reserve of millions of barrels might produce only 200 to 400 barrels per day during the best production years. Heavy oil reservoirs can be developed to significant levels of production and maintained for a period of time by supplementing natural drive force, whereas gas reservoirs normally decline more rapidly.

5.1 EXPLORATION

A basic rule of thumb in the upstream (or producing) sector of the oil and gas industry is that the best place to find new crude oil or natural gas is near where it has already been found. That is precisely what the industry does most often, for a sound business reason: the financial risk of doing so is far lower than that associated with drilling a rank wildcat hole in a prospective, but previously unproductive, area. On the other hand, there is a definite trade-off of reward for risk. The returns on drilling investment become ever leaner as more wells are drilled in a particular area because the natural distribution of oil and gas field volumes tends to be approximately log geometric. There are only a few large fields, whereas there are a great many small ones (Drew, 1997).

Exploration for hydrocarbons (oil, gas, and condensate) is commonly acknowledged to have begun with the discovery at Oil Creek, Pennsylvania, by (self-styled Colonel) Edwin Drake in 1859. This was the start of the modern global era of technology-driven advances in exploration. Traditionally, oil exploration was conducted by recognizing seeps of hydrocarbons at the surface. The Chinese, for example, used oil (mostly bitumen) obtained from seeps in medication, waterproofing, and warfare several thousand years ago. They frequently dug shallow pits or horizontal tunnels at seep locations but also, as early as 200 B.C., drilled down as much as 3,500 feet (1,067 meters) using rudimentary bamboo poles (making Drake's 69.5 ft [21.2 m] over 2,000 years later seem puny by comparison). In Baku, Azerbaijan, there are still gas and oil seeps that are permanently on fire and have been used to light caravanserai since the times of Marco Polo and the Silk Road. Similarly, seeps were recognized and exploited in the Caucasus (Groznyy region of Chechnya), Ploesti in Romania, Digboi in Assam, Sanga Sanga in eastern Borneo, and Talara in Peru.

Even Drake's well, the first to intentionally look for oil in the subsurface, was based on direct identification of seeped hydrocarbons at the surface. Initially, the oil produced was used to provide kerosene for lamps, but the later invention of automobiles drove up demand and ushered in modern methods of oil exploration. In fact, most oil until the turn of the twentieth century was in one form or another related to seep identification. However, one theory developed during this time would have a profound impact on exploration. In the mid-1800s, William Logan, the first Director of the Geological Survey of Canada, recognized oil seeps associated with the crests of convex-upward folded rocks and employed a geologist, Thomas Hunt, to formalize his anticlinal theory. This idea, however, was only recognized as a viable tool for exploration when Spindletop was discovered on the Gulf Coast of Texas in 1901. For the next 30 years, the anticlinal theory dominated exploration, to the extent that many believed that there were no other types of hydrocarbon accumulation. As a result, geologists became critical to understand the structural configurations of rock sequences which, when combined with seep occurrences, proved to be the keys to discovering the main oil-producing provinces of the United States, Mexico, and Venezuela. For a period of time before World War I, Oklahoma, Texas, and California were the world's leading production areas.

It was not until the 1920s that explorers realized that hydrocarbons could occur in situations where no anticline was preserved. For example, it was noted as far back as 1880 that oil was trapped in the Venango Sands of Pennsylvania, not in the form of an anticlinal structure, but by the lithologies occurring in a moving palaeo-shoreline. In fact, oil trapped by stratigraphy was discovered more often by chance rather than design even until the 1970s. By the 1920s, mapping of surface features was complimented by the development of seismic refraction, gravity, and magnetic geophysical methods. In particular, gravity and seismic methods proved effective in locating oil trapped against buried salt domes in the onshore Gulf of Mexico. Around this time, another significant advance in exploration of the subsurface took place with the application of geophysical techniques by the Schlumberger brothers to the measurement of properties of rocks and fluids encountered while drilling for hydrocarbons. Initially, in France in 1927, they measured the resistivity of the rocks in shallow wells (drilled primarily for water distribution), but they later went on to add other electric, sonic, and radioactive logging tools. It is now even possible to log porosity, permeability, mineralogy, and fluids and to image the structures and rock types downhole. Ultimately, these developments have been some of the main reasons why Schlumberger has become one of the largest electronic companies in the world.

Around the turn of the century and up until the 1950s, the main exploration tool used for finding oil was intensive and detailed geological mapping. This was frequently done in terrain that was remote and inhospitable. The early pioneers working their way through the jungles of Burma, India (Burma Oil Company), and Borneo (Shell); the deserts of Iraq; or the mountains of Iran (the Anglo-Persian Oil Company, which became British Petroleum), would conduct detailed evaluations of the nature and distribution of rock units. These rock units represented potential reservoirs, seals (basement rock and Cap rock), and source rock, as well as frequency, orientation, and geological history of folds or faults that could act as traps for the migrating hydrocarbons.

Following World War II, when low-cost, rapid reconnaissance of large areas became feasible, aerial remote sensing for features favored for hydrocarbon accumulation became an important and effective technique, particularly in areas of sparse vegetation cover,. Large-scale features such as faults and folds could be identified and targeted for detailed seismic acquisition. In the 1970s, this capability was improved dramatically by the use of satellite remote sensing technologies (LANDSAT).

From the 1940s to the 1960s, there were important developments in the understanding of the controls on lateral and vertical variations within reservoir sequences. In particular, the new discipline of sedimentology used modern depositional analogues from around the world to understand the nature, distribution, and controls over ancient reservoir sequences. There was also much interest generated over the discovery of carbonate oil-bearing reservoirs in West Texas and Canada (Leduc Reef) and the recognition that modern intertidal carbonate-evaporite sequences in the United Arab Emirates had equivalents in ancient reservoirs. These developments lead to the discovery of many super giant carbonate oil fields in the United States (Yates Field), Mexico (Posa Rica), Iraq (Kirkuk), and Russia (a number of Siberian oil fields).

Other tools such as geochemistry, developed during this period, have helped to quantify the level of maturity and the nature and distribution of source potential in a region. Also beginning in the 1970s, there was a significant increase in the power and reduction in the size and cost of computers, which have lead directly to a dramatic increase in the ability of geophysicists to acquire, process, and interpret large quantities of seismic data. Initially, this was in the form of 2-D reflection seismic surveys onshore. This trend has continued to the present day, and now oil companies regularly undertake mostly offshore, 3-D seismic surveys and even 4-D field surveys. Three-dimensional surveys are repeated over the same area every few years to monitor fluid movement within reservoirs and thereby optimally manage hydrocarbon recovery.

Exploration for oil and gas has progressed dramatically in the last 30 years, driven forward by the ever-increasing power and capabilities of the computer. As a result, it now takes only a fraction of the time required 20 years ago to find and develop oil fields. However, technology in itself does not find oil or gas fields; it frequently requires a flash of inspiration that is the mark of a true explorer to discover some of the major new exploration plays in such areas as Equatorial Guinea, Angola, Nigeria, Trinidad, the Gulf of Mexico, and the northern Canadian Rockies.

After an exploration effort has successfully discovered petroleum within an acceptable range of reserve potential, the challenge becomes how to best optimize extraction of recoverable reserves in a manner yielding an acceptable economic return on total cash expenditures required over the life of the project. Surface and subsurface conditions of a discovery have considerable impact on the extraction process, its related costs, and ultimate project success or failure. Technical success is one thing; economic success is another. Real world experience has shown that economic success is by far the more difficult accomplishment, as it is dependent on factors well beyond the means of science and technology.

Petroleum reserves exist as oil or gas within trapping sections of reservoir rock formed by structural and/or stratigraphic geologic features. Water is the predominant fluid found in the subsurface strata within the earth's crust. Both oil and gas have a low specific gravity relative to water and will thus float through the more porous sections of reservoir rock from their source area to the surface unless restrained by a trap. Typically, reservoir rock consists of sand, sandstone, limestone, or dolomite. A trap is a reservoir that is overlain by a dense cap rock or a zone of very low or no porosity that restrains migrating hydrocarbon. Petroleum-bearing reservoirs can exist from surface seeps to subsurface depths over four miles (6.4 kilometers) below sea level.

Reservoirs vary from being quite small to covering several thousands of acres, and they range in thickness from a few inches to hundreds of feet or more.

The process of evaluating how to best optimize extraction of recoverable reserves begins with a development plan. The development plan considers all available geologic and engineering data to make an initial estimate of reserves in place, to project recovery efficiencies and optimal recoverable reserve levels under various producing scenarios, and to evaluate development plan alternatives. Development alternatives will include the number of wells to be drilled and completed for production or injection, the well spacing and pattern, processing facility requirements, product transportation options, cost projections, project schedules, depletion plans, operational programs, and logistics and economic studies.

In general, petroleum is extracted by drilling wells from an appropriate surface configuration into the hydrocarbon-bearing reservoir or reservoirs. Wells are designed to contain and control all fluid flow at all times throughout drilling and producing operations. The number of wells required is dependent on a combination of technical and economic factors used to determine the most likely range of recoverable reserves relative to a range of potential investment alternatives.

The complexity and cost of drilling wells and installing all necessary equipment to produce reserves can vary significantly. The development of an onshore shallow gas reservoir located among other established fields may be comparatively low cost and nominally complex. A deep oil or gas reservoir located in more than 4,000 feet (1,219 meters) of water depth located miles away from other existing producing fields will push the limits of emerging technology at extreme costs. Individual wells in deep water can cost in excess of 50 million dollars to drill, complete, and connect to a producing system. Onshore developments may permit the phasing of facility investments as wells are drilled and production established to minimize economic risk. However, offshore projects may require 65% or more of the total planned investments to be made before production start up; this imposes significant economic risk.

Once production begins, the performance of each well and reservoir is monitored, and a variety of engineering techniques are used to progressively refine reserve recovery estimates over the producing life of the field. The total recoverable reserves are not known with complete certainty until the field has produced to depletion or to its economic limit and abandonment.

The ultimate recovery of the original-in-place may be as high as 33% for oil and 80% or more for gas. There are three phases of recovering reserves (Figure 5–1):

1. *Primary recovery* occurs as wells produce because of natural energy from expansion of gas and water within the producing formation, which pushes fluids into the well bore and lifts them to the surface.

2. *Secondary recovery* occurs as artificial energy is applied to inject fluids into the well bore and lift fluids to the surface. This may be accomplished by injecting gas down a hole, installing a subsurface pump, or injecting gas or water into the formation itself. Secondary recovery is done when well, reservoir, facility, and economic conditions permit.

3. *Tertiary recovery* (*enhanced recovery*) occurs when means of increasing fluid mobility within the reservoir are introduced in addition to secondary techniques. This may be accomplished by introducing additional heat into the formation to lower the viscosity (thin the oil) and improve its ability to flow to the well bore. Heat may be introduced by either injecting steam in a *steam flood* or injecting oxygen to enable the ignition and combustion of oil within the reservoir in a *fire flood*. Such methods are undertaken only in a few unique situations where technical, environmental, and economic conditions permit. Most gas reserves are produced during the primary recovery phase. Secondary recovery has significantly contributed to increasing oil recovery.

Many technical assumptions become better understood and more certain with the evaluation of performance data over the producing life of a field. One of the most critical assumptions, however, remains uncertain and holds project success at risk to the very end—the oil and gas price forecast.

Crude oil development and production can include up to three distinct phases: primary, secondary, and tertiary (or enhanced) recovery. During primary recovery, the natural pressure of the reservoir or gravity drive oil into the well bore, and artificial lift techniques (such as pumps) bring the oil to the surface. Typically, only about 10% of a

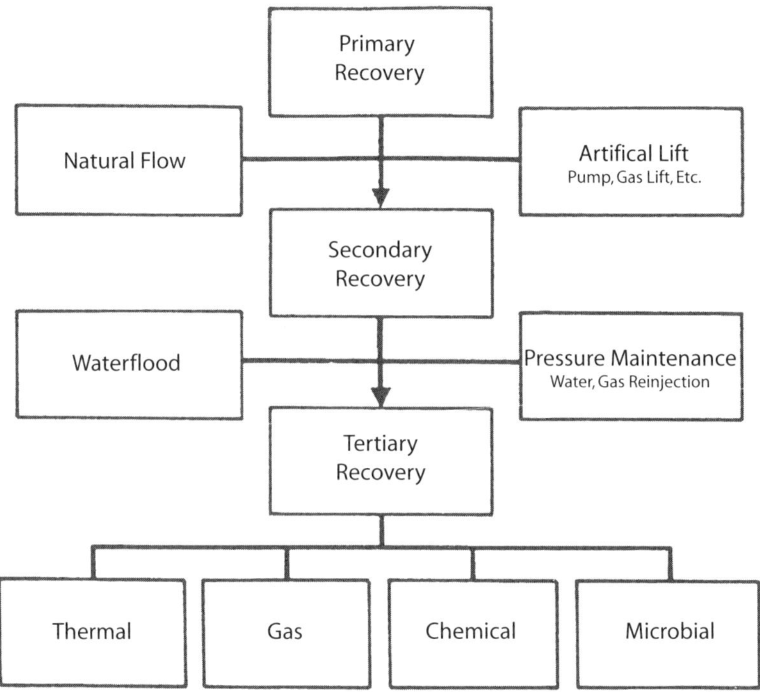

Figure 5–1 *Methods for oil recovery.*

reservoir's original oil in place is produced during primary recovery. Secondary recovery techniques added to the field's productive life, generally injecting water or gas to displace oil and drive it to a production well bore, result in the recovery of an additional 20% to 40% of the original oil in place.

Usually, the first stage in the extraction of crude oil is to drill a well into the underground reservoir. Often, many wells (*multilateral wells*) will be drilled into the same reservoir to ensure that the extraction rate will be economically viable. Also, some wells (*secondary wells*) may be used to pump water, steam, acids, or various gas mixtures into the reservoir to raise or maintain the reservoir pressure and so maintain an economic extraction rate.

If the underground pressure in the oil reservoir is sufficient, the oil will be forced to the surface under this pressure (primary recovery). Natural gas (*associated natural gas*) is often present, which also supplies needed underground pressure. In this situation, it is sufficient to place an arrangement of valves (referred to as the *Christmas tree* because of the multiple branches) on the wellhead to connect the well to a pipeline network for storage and processing.

Over the lifetime of the well, the pressure will fall, and at some point, there will be insufficient underground pressure to force the oil to the surface. Secondary oil recovery uses various techniques to aid in recovering oil from depleted or low-pressure reservoirs. Sometimes, pumps, such as beam pumps (*horsehead pumps*) and electrical submersible pumps, are used to bring the oil to the surface. Other secondary recovery techniques increase the reservoir's pressure by water injection; natural gas reinjection; and gas lift, which injects air, carbon dioxide, or some other gas into the reservoir.

Tertiary oil recovery methods take oil recovery one step further (Chakma et al., 1991) and rely on methods that reduce the viscosity of the oil and increase oil mobility, compared to the natural- or induced-energy methods of primary and secondary recovery. Tertiary recovery is started *before* secondary recovery techniques are no longer enough to sustain production. For example, thermally enhanced oil recovery methods are recovery methods in which the oil is heated to make it easier to extract; usually steam is used for heating the oil.

Conventional primary and secondary recovery processes are ultimately expected to produce about one-third of the original oil discovered, although recoveries from individual reservoirs can range from less than 5% to as high as 80% of the original oil-in-place. This broad range of recovery efficiency is a result of variations in the properties of the specific rock and fluids involved from reservoir to reservoir as well as the kind and level of energy that drives the oil to producing wells, where it is captured.

Conventional oil production methods may be unsuccessful because the management of the reservoir was poor or because reservoir heterogeneity prevented the recovery of crude oil in an economical manner. Reservoir heterogeneity, such as fractures and faults, can cause reservoirs to drain inefficiently by conventional methods. Also, highly cemented or shale zones can produce barriers to the flow of fluids in reservoirs and lead to high residual oil saturation. Reservoirs containing crude oils with low API gravity often cannot be produced efficiently without application of *enhanced oil recovery* (EOR) methods because of the high viscosity of the crude oil. In some cases, the reservoir pressure is depleted prematurely by poor reservoir management practices that create reservoirs with low energy and high oil saturation.

As might be expected, the type of exploration technique employed depends upon the nature of the site; it is site specific. Similarly, the recovery techniques applied to a particular well or field are *site specific* (as they are for many environmental operations). For example, in areas where little is known about the subsurface, preliminary reconnaissance techniques are necessary to identify potential reservoir systems that warrant further investigation. Reconnaissance techniques that have been employed to make inferences about the subsurface structure include *satellite and high-altitude imagery* and *magnetic and gravity surveys*.

Once an area has been selected for further investigation, more detailed methods (such as the *seismic reflection* method) are brought into play. Drilling is the final stage of the exploratory program and is in fact the only method by which a petroleum reservoir can be conclusively identified. However, in keeping with the concept of site specificity, drilling may be the only option in some areas for commencement of the exploration program. The risk involved in the drilling operation depends upon previous knowledge of the site subsurface. Thus, there is the need to relate the character of the exploratory wells at a given site to the characteristics of the reservoir.

As mentioned, there are several methods by which recovery can be achieved, ranging from methods in which recovery is due to reservoir energy to methods in which considerable energy must be added to the reservoir to produce the oil. The effect of the method on the oil and on the reservoir must be considered before application.

This chapter, for the most part, deals with those methods that are applied for the recovery of conventional crude oil and, in some cases, for the recovery of heavy oil. Methods that are being proposed for the recovery of bitumen from tar deposits are presented elsewhere.

Once the well is completed, the flow of oil into the well is commenced. For limestone reservoir rock, acid is pumped down the well and out the perforations. The acid creates channels in the limestone that lead oil into the well. For sandstone reservoir rock, a specially blended fluid containing *proppants* (sand, walnut shells, aluminum pellets) is pumped down the well and out the perforations. The pressure from this fluid makes small fractures in the sandstone that allow oil to flow into the well, while the proppants hold these fractures open. In most wells, acidizing or fracturing the well starts the oil flow. Once the oil is flowing, the oil rig is removed from the site, and production equipment is set up to extract the oil from the well.

A well is always carefully controlled in its flush stage of production to prevent the potentially dangerous and wasteful *gusher*. This condition is (hopefully) prevented by the blowout preventer and the pressure of the drilling mud.

As already noted, crude oil accumulates over geological time in porous underground rock formations called reservoirs that are at varying depths in the earth's crust; in many cases, elaborate, expensive equipment is required to get it from there. The oil is usually found trapped in a layer of porous sandstone, which lies just beneath a dome-shaped or folded layer of some nonporous rock, such as limestone. In other formation, the oil is trapped at a fault, or break in the layers of the crust.

Generally, crude oil reservoirs exist with an overlying *gas cap*, in communication with aquifers, or both. The oil resides together with water and free gas in very small holes (pore spaces) and fractures. The size, shape, and degree of interconnection of the pores vary considerably from place to place in an individual reservoir. Below the oil layer, the sandstone is usually saturated with salt water. The oil is released from this formation by drilling a well and puncturing the limestone layer on either side of the limestone dome or fold. If the peak of the formation is tapped, only the gas is obtained. If the penetration is made too far from the center, only salt water is obtained.

The oil in such formation is usually under such great pressure that it flows naturally, and sometimes with great force, from the well. However, in some cases, this pressure later diminishes so that the oil must be pumped from the well. Natural gas or water is sometimes pumped into the well to replace the oil that is withdrawn. This is called repressurizing the oil well.

The anatomy of a reservoir is complex and site specific, microscopically and macroscopically. Because of the various types of accumulations and the wide ranges of both rock and fluid properties, reservoirs respond differently and must be treated individually.

Conventional crude oils are brownish green to black liquids of specific gravity in a range from about 0.810 to 0.985 that have a boiling range from about 20°C (68°F) to above 350°C (660°F), above which active decomposition ensues when distillation is attempted. The oils contain 0% to 35% or more of gasoline, as well as varying proportions of kerosene hydrocarbons and higher-boiling constituents up to the viscous and nonvolatile compounds present in lubricant oil and in asphalt. The composition of the crude oil obtained from a well is variable and depends not only on the original composition of the oil in situ but also on the manner of production and the stage reached in the life of the well or reservoir.

For a newly opened formation and under ideal conditions, the proportions of gas may be so high that the oil is, in fact, a solution of liquid in gas that leaves the reservoir rock so efficiently that a core sample will not show any obvious oil content. A general rough indication of this situation is a high ratio of gas to oil produced. This ratio may be zero for fields in which the rock pressure has been dissipated. The oil must be pumped out to as much as 50,000 ft^3 or more of gas per barrel of oil in the so-called condensate reservoirs (in which a very light crude oil [0.80 specific gravity or lighter] exists as vapor at high pressure and elevated temperature).

New methods to drill for oil are continually being sought. These include using directional or horizontal drilling techniques to reach oil under ecologically sensitive areas and using lasers to drill oil wells.

Directional drilling is also used to reach formations and targets not directly below the penetration point or to drill from shore to locations under water (Figure 5–2). A controlled deviation may also be used from a selected depth in an existing hole to attain economy in drilling costs. Various types of tools are used in directional drilling, along with instruments to help orient their position and measure the degree and direction of deviation; two such tools are the *whipstock* and the *knuckle joint*. The whipstock is a gradually tapered wedge with a chisel-shaped base that prevents rotation after it has been forced into the bottom of an open hole. As the bit moves down, it is deflected by the taper about five degrees from the alignment of the existing hole.

Approximately one-third of the world's crude oil is produced from offshore fields, usually from steel drilling platforms set on the ocean floor. In shallow, calm waters, these may be little more than a well-

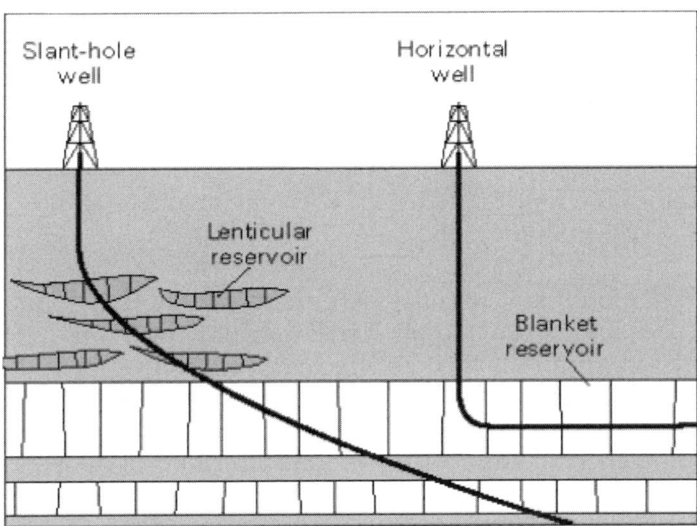

Figure 5–2 *Directional drilling.*

head and workspace. The larger ocean rigs, however, include not only the well equipment but also processing equipment and extensive crew quarters. Recent developments in ocean drilling include the use of floating tension leg platforms that are tied to the sea floor by giant cables and drill ships, which can hold a steady position above a seafloor well using constant, computer-controlled adjustments. Subsea satellite platforms, where all of the necessary equipment is located on the ocean bed at the well site, have been used for small fields located in producing areas. In Arctic areas, islands are built from dredged gravel and sand to provide platforms capable of resisting drifting ice fields.

Drilling does not end when production commences. Extension wells must be drilled to define the boundaries of the crude oil pool. In-field wells are necessary to increase recovery rates, and service wells are used to reopen wells that have become clogged. Additionally, wells are often drilled at the same location but to different depths to test other geological structures for the presence of crude oil.

Once the final depth has been reached, the well is completed to allow oil to flow into the casing in a controlled manner. First, a *perforating gun* is lowered into the well to the production depth. The gun has explosive charges that create holes in the casing through which oil can flow. After the casing has been perforated, a small-diameter pipe (*tubing*) is run into the hole as a conduit for oil and gas to flow up the well, and a *packer* is run down the outside of the tubing. When the packer is set at the production level, it is expanded to form a seal around the outside of the tubing. Finally, a multivalve structure (the *Christmas tree*; Figure 5–3) is installed at the top of the tubing and cemented to the top of the casing. The Christmas tree allows control the flow of oil from the well.

Tight formations are occasionally encountered, and it becomes necessary to encourage flow. Several methods are used, one of which involves setting off small explosions to fracture the rock. In sandstone, the preferred method is hydraulic fracturing. If the formation is mainly limestone, hydrochloric acid is sent down the hole to create channels in the rock. The acid is inhibited to protect the steel casing.

Finally, the drilling job is complete when the drill bit penetrates the reservoir. The reservoir is evaluated to see whether the well represents the discovery of a *prospect* or it is a dry hole.

Evaluation is usually initiated by examining the cuttings from the well bore for evidence of hydrocarbons. The evaluation of these cuttings helps pinpoint the possible producing intervals in the well bore. At this time, a wire line is lowered into the hole and an electric log is run to help ascertain possible producing intervals, presence of hydrocarbons, and detailed information about the different formations throughout the well bore. Further tests can also be run on individual formations within the well bore, such as pressure tests, formation fluid recovery, and sidewall core analysis. On the other hand, if the hole is dry, it is plugged and abandoned.

If hydrocarbons are detected, the completion process begins. The only thing visible at the wellhead after the drill rig leaves the site is a series of valves and gauges connected vertically to each other and attached to the top of the well (the Christmas tree). This allows the amount of hydrocarbons to flow from the well, and it prevents leakage at the surface. Hydrocarbons come in varying densities and viscosities; reservoir traps also have variations in porosity, permeability, pressures, and temperatures. These factors exert an influence on how easily the oil can be removed from the reservoir.

It is the differential pressure between the reservoir and the open hole that moves the hydrocarbons out of the reservoir, into the well, and

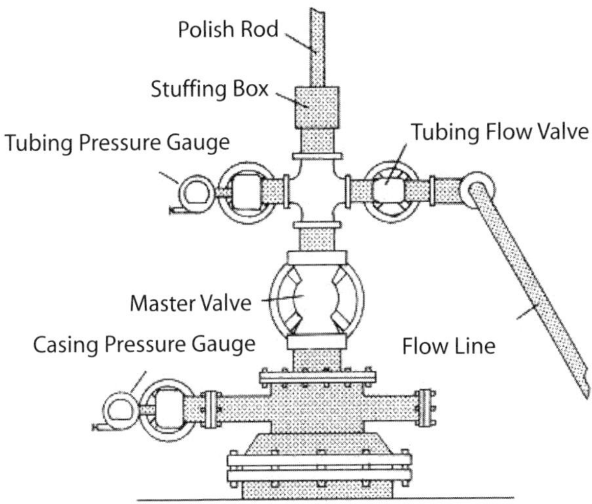

Figure 5–3 *The Christmas Tree.*

up to the surface. The pressure may be the result of a number of forces. For example, water located below the oil layer may be pressing upward; when this occurs, it is referred to as a water drive system. If the gas cap located above the oil is causing a downward pressure, it is referred to as a gas-cap drive system.

5.2 Primary Recovery (Natural) Methods

In most reservoir traps, initial pressure (reservoir energy) is sufficient to push the oil to the surface of the production well with only minimal help from a downhole pump. With declining well pressures, however, it becomes more difficult to get the hydrocarbons to the surface. Sometimes, artificial oil lift is needed. The most common installation uses a pump at the bottom of the production tubing that is operated by a motor and a *walking beam* (an arm that rises and falls like a seesaw) on the surface. A string of solid metal *sucker rods* connects the walking beam to the piston of the pump. Another method, called gas lift, uses gas bubbles to lower the density of the oil, allowing the reservoir pressure to push it to the surface. Usually, the gas is injected down the annulus between the casing and production tubing and through a special valve at the bottom of the tubing. In a third type of artificial lift, produced oil is forced down the well at high pressure to operate a pump at the bottom of the well.

With the artificial lift methods described above, oil may be produced as long as there is enough nearby reservoir pressure to create flow into the well bore. Inevitably, however, a point is reached at which commercial quantities no longer flow into the well. In most cases, only a minority of the amount of oil originally present can be produced by naturally occurring reservoir pressure alone, and in some cases (e.g., where the oil is quite viscous and at shallow depths), primary production is not economically possible at all.

Crude oil moves out of the reservoir into the well by one or more of three processes. These processes are *dissolved gas drive, gas cap drive,* and *water drive.* Early recognition of the type(s) of drive involved is essential to the efficient development of an oil field.

In *dissolved gas drive* (*solution gas drive*) (Figure 5–4) the propulsive force is the gas in solution in the oil, which tends to come out of solution because of the pressure release at the point of penetration of a well. Dissolved gas drive is the least efficient type of natural drive as it is difficult to control the gas-oil ratio. The bottom-hole pressure drops rapidly, and the total eventual recovery of petroleum from the reservoir may be less than 20%.

If gas overlies the oil beneath the top of the trap, it is compressed and can be utilized to drive the oil into wells situated at the bottom of the oil-bearing zone. This is the *gas cap drive* (Figure 5–5). By producing oil only from below the gas cap, it is possible to maintain a high gas-oil ratio in the reservoir until almost the very end of the life of the pool. If, however, the oil deposit is not systematically developed so that bypassing of the gas occurs, an undue proportion of oil is left behind. The usually recovery of petroleum from a reservoir in a gas cap field is 40% to 50%.

Usually the gas in a gas cap (*associated natural gas*) contains methane and other hydrocarbons that may be separated out by compressing the gas. A well-known example is *natural gasoline,* which was formerly referred to as *casinghead gasoline* or *natural gas gasoline.* However, at high pressures, such as those existing in the deeper fields, the density of the gas increases and the density of the oil decreases until they form a single phase in the reservoir. These are the so-called retrograde condensate pools because a decrease (instead of an increase) in pres-

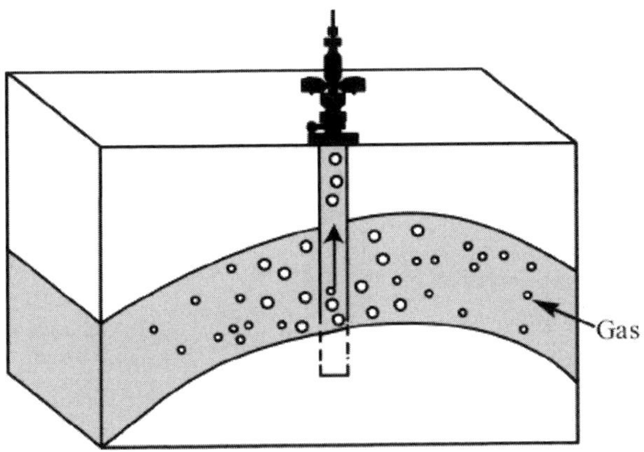

Figure 5–4 *Solution-gas drive.*

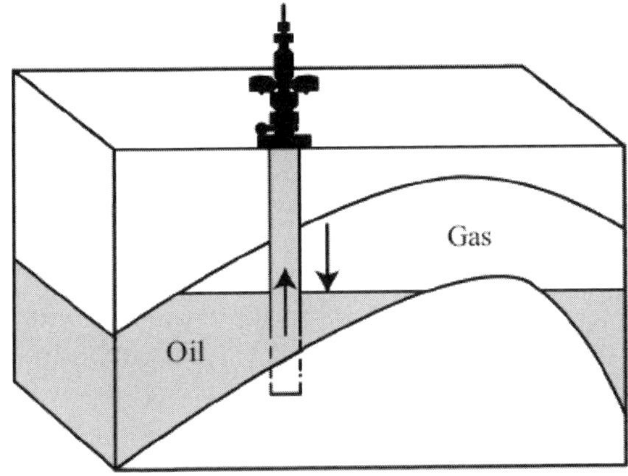

Figure 5–5 *Gas-cap drive.*

sure brings about condensation of the liquid hydrocarbons. When this reservoir fluid is brought to the surface and the condensate is removed, a large volume of residual gas remains. The modern practice is to cycle this gas by compressing it and inject it back into the reservoir, thus maintaining adequate pressure within the gas cap and preventing condensation in the reservoir. Such condensation prevents recovery of the oil, for the low percentage of liquid saturation in the reservoir precludes effective flow.

The most efficient propulsive force in driving oil into a well is natural *water drive*, in which the pressure of the water forces the lighter recoverable oil out of the reservoir (Figure 5–6). In anticlinal accumulations, the structurally lowest wells around the flanks of the dome are the first to come into water. Then the oil-water contact plane moves upward until only the wells at the top of the anticline are still producing oil; eventually these also must be abandoned as the water displaces the oil.

In a water drive field it is essential that the removal rate be adjusted so that the water moves up evenly as space is made available for it by the removal of the hydrocarbons. An appreciable decline in bottom-hole pressure is necessary to provide the pressure gradient required to cause water influx. The pressure differential needed depends on the reservoir permeability; the greater the permeability, the less the difference in pressure necessary. The recovery of petroleum from the reservoir in properly operated water drive pools may run as high as 80%. The force behind the water drive may be hydrostatic pressure, the expansion of the reservoir water, or a combination of both. Water drive is also used in certain submarine fields.

Gravity drive is an important factor when oil columns of several thousands of feet exist, as they do in some North American fields. Furthermore, the last bit of recoverable oil is produced in many pools by gravity drainage of the reservoir. Another source of energy during the early stages of withdrawal from a reservoir containing undersaturated oil is the expansion of that oil as the pressure reduction brings the oil to the bubble point (the pressure and temperature at which the gas starts to come out of solution).

For primary recovery operations, it is sufficient to force the oil to the surface but if the reservoir energy is not sufficient to force the oil to the surface, however the oil must be pumped to the surface. Even so, nothing is added to the reservoir to increase or maintain the reservoir energy or to sweep the oil toward the well. The rate of production from a flowing well tends to decline as the natural reservoir energy is expended. When a flowing well is no longer producing at an efficient rate, a pump is installed.

The recovery efficiency for primary production is generally low when liquid expansion and solution gas evolution are the driving mechanisms. Much higher recoveries are associated with reservoirs with water and gas cap drives and with reservoirs in which gravity effec-

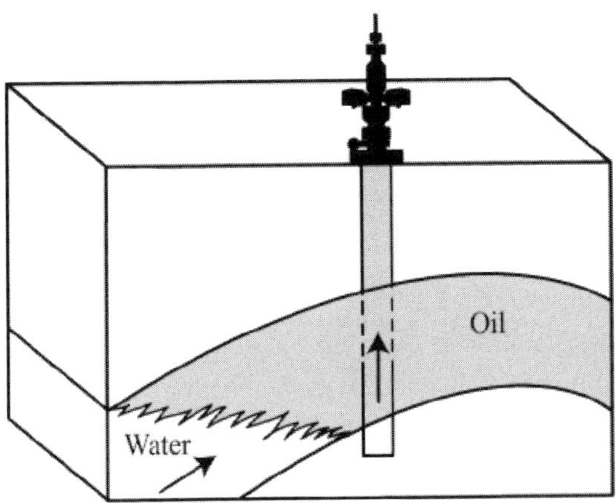

Figure 5–6 *Water drive.*

tively promotes drainage of the oil from the rock pores. The overall recovery efficiency is related to how the reservoir is delineated by production wells. For maximum recovery by primary recovery, it is often preferable to sink several wells into a reservoir, thereby bringing about recovery by a combination of the methods outlined here.

When production begins to drop off, it may be time for the well to receive a work over (a major repairing and cleaning out of all pipes). Producing wells are like anything else; they require periodic maintenance. Corrosion can roughen pipe walls or cause failure, allowing product to leak onto the surface. Pieces of rock from the side of the well may break off and fall into the well, clogging it. Natural gas pipes tend to accumulate paraffin (hydrate compounds that build up inside the pipe causing restrictions). Maintenance can include everything from cleaning fluids being injected into the pipes to wire brushes being inserted to brush the pipes clean. Residues are flushed from the system before it is reconnected.

But work over is not restricted only to the hardware; it may also be applied to the downhole portion of the rock formation. Often, the formation through which the hydrocarbons are flowing becomes clogged, which diminishes the volume of product reaching the well. Two processes used to improve formation characteristics are acidizing and fracturing. Acidizing involves injecting an acid into a soluble formation, such as a carbonate, where it dissolves rock. This process enlarges the existing voids and increases permeability. Hydraulic fracturing (*fracking*) involves injecting a fluid into the formation under significant pressure. This makes existing small fractures larger and creates new fractures.

In conventional oil production, the concept of applying more than one recovery technology, one after the other, to a reservoir is well established. When primary production declines and becomes less economic, producers investigate the opportunity to water flood the reservoir as a secondary recovery technique. Finally, tertiary methods may be applied when water floods yield diminishing returns. Heavy oil and oil sands have a shorter history, and generally reservoirs have been subject to only one recovery technology. In the case of oil sands, primary and secondary recovery technologies (as defined for conventional oil) are not applicable because bitumen is not mobile at reservoir conditions. Therefore, oil sands developments generally start with a thermal recovery technology, which would be considered a tertiary or enhanced recovery method for conventional oil. However, as the development of heavy oil reservoirs and oil sand deposits matures, the concept of applying more than one recovery technology in a specific order is likely to also be applied to heavy oil and bitumen reservoirs. In particular, in the Lloydminster area, researchers and producers have already been investigating for several years the concept of follow-up recovery technologies since primary production is not usually feasible or economic for heavy oil or tar sand bitumen

5.3 Secondary Recovery

Over the lifetime of the well, the pressure will fall, and at some point, there will be insufficient underground pressure to force the oil to the surface. If economical (as it often is) the remaining oil in the well is extracted using secondary oil recovery methods.

When a large part of the crude oil in a reservoir cannot be recovered by primary means, a method for supplying extra energy must be chosen. Most often, secondary recovery is accomplished by injecting gas or water into the reservoir to replace produced fluids and thus maintain or increase the reservoir pressure. When gas alone is injected, it is usually put into the top of the reservoir, where petroleum gases normally collect to form a gas cap. Gas injection can be a very effective recovery method in reservoirs where the oil is able to flow freely to the bottom by gravity. When this gravity segregation does not occur, however, other means must be sought.

Secondary oil recovery methods use various techniques to aid in recovering oil from depleted or low-pressure reservoirs. Sometimes, pumps on the surface or submerged (electrical submersible pumps, ESPs) are used to bring the oil to the surface. Other secondary recovery techniques increase the reservoir's pressure by water injection and gas injection, which inject air or some other gas into the reservoir.

Together, primary recovery and secondary recovery allow 25% to 35% of the reservoir's oil to be recovered. Primary (or conventional) recovery can leave as much as 70% of the petroleum in the reservoir. Such effects as microscopic trapping and by-passing are the more obvious reasons for the low recovery.

There are two main objectives in secondary crude oil production. One objective is to supplement the depleted reservoir energy pressure, and the other is to sweep the crude oil from the injection well toward and into the production well. Essentially, secondary oil recovery involves the introduction of energy into a reservoir to produce more oil. For example, the addition of materials to reduce the interfacial tension of the oil results in a higher recovery of oil.

The most common follow-up (or secondary recovery) operations involve the application of pumping operations or the injection of materials into a well to encourage movement and recovery of the remaining petroleum. The pump, generally known as the *horsehead pump* (*pump jack*, *nodding donkey*, or *sucker rod pump*) (Figure 5–7), provides mechanical lift to the fluids in the reservoir.

The pump is powered by an electric motor, although some older, less producing wells use propane as an alternative source of power rather than a large power grid. Some of these wells can even use the natural gas from the casing as fuel, and the well can be completely self sufficient. The engine of the pump runs a set of pulleys to the transmission, which in turn drives a pair of cranks that generally have counter weights on them to assist the motor in lifting the heavy string of sucker rods. The cranks in turn raise and lower one end of the "I" beam which is free to move on an "A" frame. A metal or fiberglass cable called a bridle, connects the horsehead to the polished rod.

The bridle follows the curve of the horsehead as it lowers and rises to create a completely vertical stroke. The up-and-down movement of the sucker rods forces the oil up the tubing to the surface. A walking beam powered by a nearby engine may supply this vertical movement,

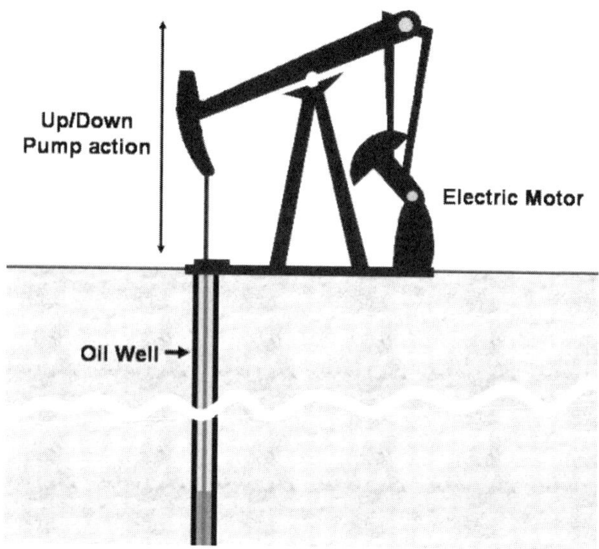

Figure 5–7 *A horsehead pump.*

or it may be brought about through the use of a pump jack, which is connected to a central power source by means of pull rods. Depending on the size of the pump, it generally produces up to one third of a barrel of an oil-water emulsion at each stroke. The size of the pump is also determined by the depth and weight of the oil to be removed, with deeper extraction requiring more power to move the heavier lengths of polish rod.

Electrically powered centrifugal pumps and submersible pumps (both pump and motor are in the well at the bottom of the tubing) have also proven their production capabilities in numerous applications.

Other *secondary oil recovery* operations involve the injection of water or gas into the reservoir. When water is used, the process is called a *waterflood*; with gas, a *gasflood*. Separate wells are usually used for injection and production. The injected fluids maintain reservoir pressure or repressure the reservoir after primary depletion and displace a portion of the remaining crude oil to production wells. In fact, the first method recommended for improving the recovery of oil was probably the reinjection of natural gas; there are indications that gas injection was utilized for this purpose before 1900 (Craft and Hawkins, 1959; Frick, 1962). These early practices were implemented to increase the immediate productivity, and are therefore classified as pressure maintenance projects. Recent gas injection techniques have been devised to increase the ultimate recovery, thus qualifying them as secondary recovery projects.

Water injection often increases oil recovery to twice that expected from primary means alone. The wells to be used for injecting water are usually located in a pattern that will best push oil toward the production wells. Some oil reservoirs (the East Texas field, for example) are connected to large, active water reservoirs, or aquifers, in the same formation. In such cases, it is necessary only to reinject water into the aquifer in order to help maintain reservoir pressure.

Using techniques such as gas and water injection, there is no change in the state of oil. Similarly, there is no change in the state of the oil when using miscible fluid displacement technologies. The analogy that might be used is that of a swimmer (in water); there is no change to the natural state of the human body when the swimmer is moved by the water.

The success of secondary recovery processes depends on the mechanism by which the injected fluid displaces the oil (displacement efficiency) and on the volume of the reservoir that the injected fluid enters (conformance or sweep efficiency). In most proposed secondary projects, water does both these things more effectively than gas. It must be decided if the use of gas offers any economic advantages because of availability and relative ease of injection. In reservoirs with high permeability and high vertical span, the injection of gas may result in high recovery factors as a result of gravity segregation, as described in a later section. On the other hand, if the reservoir lacks either adequate vertical permeability or the possibility for gravity segregation, a frontal drive similar to that used for water injection can be used (dispersed gas injection). Dispersed gas injection is anticipated to be more effective in reservoirs that are relatively thin and have little dip. Injection into the top of the formation (or into the gas cap) is more successful in reservoirs with higher vertical permeability (200 millidarcies or more) and enough vertical relief to allow the gas cap to displace the oil downward.

Vaporization is another recovery mechanism used to inject gas into oil reservoirs. A portion of the oil affected by the *dry* injection gas is vaporized into the oil and transported to the production wells in the vapor phase. In some instances, this mechanism has been responsible for a substantial amount of the secondary oil produced.

During the withdrawal of fluids from a well, it is usual practice to maintain pressures in the reservoir at or near the original levels by pumping either gas or water into the reservoir as the hydrocarbons are withdrawn. This practice has the advantages of retarding the decline in the production of individual wells and considerably increasing the ultimate yield. It also may conserve gas that otherwise would be wasted and reduce the disposal of brines that otherwise might pollute surface and near-surface potable waters. In older fields, it was not the usual practice to maintain the reservoir pressure, and it is now necessary to obtain petroleum from these fields by means of secondary recovery projects.

Considerable experimentation has been carried out on the use of different types of input gas. Examples are *wet casinghead gas*; *enriched gas*; *liquefied petroleum gas* (*LPG*), such as butane and propane; high-pressure gas; and even nitrogen. High-pressure gas not only pushes oil through the reservoir but may also produce a hydrocarbon exchange that increases the concentration of liquid petroleum gases in the oil.

Waterflooding is a form of oil recovery wherein some of the energy required to move the oil from the reservoir rock into a producing well is supplied from the surface by means of water injection.

While the above numbers are typical, it should be noted that the ratio of waterflood oil to primary oil varies greatly with depth. Shallow reservoirs tend to have both low pressures and small amounts of dissolved gas. This low available drive energy often translates into low primary recovery, sometimes less than 5%. These reservoirs often recover much more secondary oil than primary oil.

Conversely, deeper oil reservoirs tend to have higher pressures, more dissolved gas, and consequently, better recoveries on primary production. In a 10000-foot oil reservoir with a gas cap and high solution gas, 40% primary recovery is possible. This obviously leaves only 10% likely to be recovered by waterflooding, giving a secondary to primary ratio of 1:4.

Generally, the selection of an appropriate flooding pattern for waterflooding the reservoir depends on the quantity and location of accessible wells. Frequently, producing wells can be converted to injection wells. In other circumstances, it may be necessary or advantageous to drill new injection wells.

In designing a waterflood project, it is general practice to locate injection and producing wells in a regular geometric pattern so that a symmetrical and interconnected network is formed. However, the relative location of injectors and producers depends on (1) reservoir geometry, (2) lithology, (3) reservoir depth, (4) porosity, (5) permeability, (6) continuity of reservoir rock properties, (7) magnitude and distribution of fluid saturations, and last but certainly not least, (8) fluid (oil) properties. Overall, the goal is to increase the mobility of the oil.

The mobility of oil is the effective permeability of the rock to the oil divided by the viscosity of the oil:

$$\lambda = k / \mu$$

where λ is the mobility, *md/cp*, k is the effective permeability of reservoir rock to a given fluid, *md*, and μ is the fluid viscosity, *cp*. The mobility ratio (M) is the mobility of the water divided by the mobility of oil:

$$M = K_{rw}\mu_o / K_{ro}\mu_w$$

where K_{rw} is the relative permeability to water, K_{ro} is the relative permeability to oil, μ_o is the viscosity of the oil, and μ_w if the viscosity of water.

The mobility ratio (M) infers that K_o is the mobility of oil ahead of the front (measured at S_{wc}), and K_w is the mobility of water at average water saturation in the water-contacted portion of the reservoir.

The mobility ratio of a waterflood will remain constant before breakthrough, and it will increase after breakthrough, corresponding to the increase in water saturation and relative permeability to water in the water-contacted portion of the reservoir. The mobility ratio at water breakthrough is the term that is of significance in describing the relative mobility ratio. A value $M < 1$ indicates a favorable displacement as oil moves faster than water, and

$M = 1$ indicates a favorable displacement as both oil and water move at equal speed. A value $M > 1$, on the other hand. indicates an unfavorable displacement as water moves faster than oil.

Whenever it is feasible, the injection design should take advantage of gravity, that is to say, dipping or inclined reservoirs, underlying aquifers, or gas caps. A variety of injection-production patterns have been employed in waterflooding operations. Pattern selection is important because it can affect the area swept by the injected fluid. Actual well locations frequently do not coincide with the ideal well locations associated with the well patterns used.

Generally, the choice of pattern (Figure 5–8) for waterflooding must be consistent with the existing wells. The objective is to select the proper pattern that will provide the injection fluid with the maximum possible contact with the crude oil to minimize bypassing by the water.

In a *four-spot pattern,* the distance between all like wells is constant. Any three injection wells form an equilateral triangle with a production well at the center. The four-spot pattern may be used when the injectivity is high or the heterogeneity is minimal.

In a *five-spot pattern,* the distance between all like wells is constant. Four injection wells form a square with a production well at the center. If existing wells were drilled on square patterns, five-spot patterns (as well as nine-spot patterns) are most commonly used since they allow easy conversion to a five-spot waterflood. If the injectivity is low or the heterogeneity is large, the five-spot pattern may not be the best choice, and a nine-spot may be more efficient.

In the *seven-spot pattern,* the injection wells are located at the corner of a hexagon with a production well at its center. If the reservoir characteristics yield lower than preferred injection rates, either a seven-spot or a nine-spot pattern should be considered because there are more injection wells per pattern than producing wells.

In the *nine-spot pattern,* the arrangement is similar to that of the five-spot pattern but with an extra injection well drilled at the middle of each side of the square. The pattern essentially contains eight injectors surrounding one producer. If existing wells were drilled on square patterns, nine-spot patterns (as well as five-spot patterns) are most commonly used. If the reservoir characteristics yield lower injection rates than those desired, one should consider using either a nine-spot

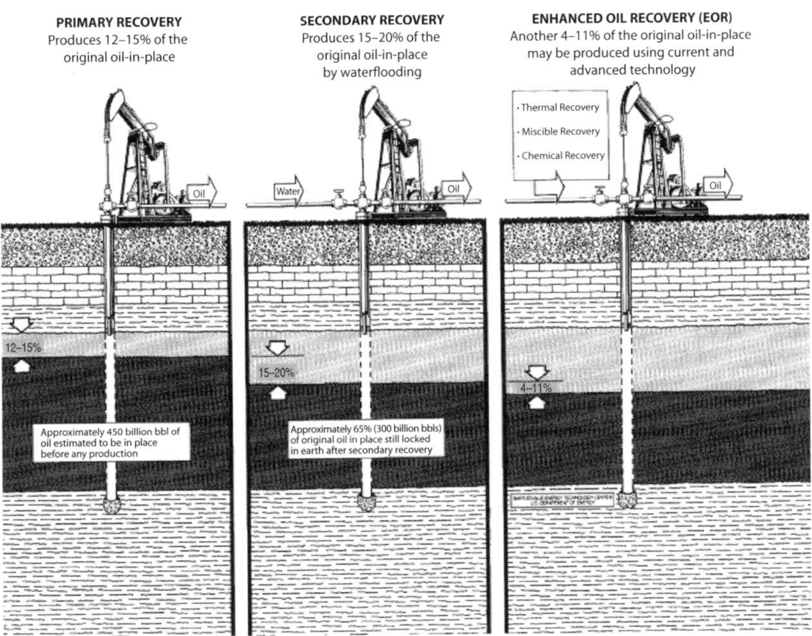

Figure 5–8 *Oil production methods.*

pattern or a seven-spot pattern where there are more injection wells per pattern than producing wells.

In the *inverted seven-spot pattern,* the arrangement is similar to the normal seven-spot pattern except that there is now an injector well where there was a producer well in the normal seven-spot pattern. Likewise, there are now producer wells where the injector wells were in the normal seven-spot pattern. The inverted seven-spot pattern may be used when the injectivity is high or the heterogeneity is minimal.

In the *inverted nine-spot pattern,* the arrangement of the wells is similar to the normal nine-spot pattern except the position of the producer well in the normal nine-spot pattern is occupied by an injector well. Likewise, where the positions of the injector wells were in the normal nine-spot, there are now producer wells. If the reservoir is fairly homogenous and the mobility ratio is unfavorable, the inverted nine-spot pattern may be promising.

In the *direct line-drive pattern,* the lines of injection and production are directly opposite to each other. If the injectivity is low or the heterogeneity is large, direct line drive is a good option. Anisotropic permeability, permeability trends, and oriented fracture systems favor line drive patterns.

In the *staggered line-drive pattern,* the wells are in lines as in the direct line-drive pattern, but the injectors and producers are no longer directly opposed. The injectors and producers are laterally displaced by a specified distance that is dependent upon the distance between wells of the same type and the distance between the lines of injector wells and producer wells. The staggered line-drive pattern is also effective for reservoirs where there is anisotropic permeability or where permeability trends or oriented fracture systems are operational.

Reservoir uniformities also dictate the choice of pattern, and mobility ratio has an important influence as well. If the ratio is unfavorable, the injectivity of an injector will exceed the productivity of a producer, and water injection will supersede oil production. To balance the production with the water injection, more producers than injectors are required. On the other hand, if the mobility ratio is favorable, the injectivity is impaired, and the pattern should have more injectors than producers. Consideration should be given to the desired oil production level after waterflooding, which will influence the ratio of injectors to producers (Table 5–1) to aid in increasing the areal sweep efficiency.

Table 5–1 Ratio of Injectors to Producers for Various Well Patterns

Pattern	Ratio of Producing Wells to Injection Wells	Drilling Pattern Required
Four—Spot	2	Equilateral triangle
Five—Spot	1	Square
Seven—Spot	1/2	Equilateral triangle
Inverted Seven—Spot	2	Equilateral triangle
Nine—Spot	1/3	Square
Inverted Nine—Spot	3	Square
Direct line drive	1	Rectangle
Staggered line drive	1	Offset lines of wells

In the ideal case, the displacement of oil by water takes place in the form of a homogenous front and, depending on the mobility ratio of oil and water, the front starts to generate fingers resulting in bypasses of the oil. After the waterfront breaks through at the producer, high water cut and low oil production characterizes the production behavior of this producer. The velocity of the frontal movement depends on the amount of water injected in the wells surrounding the producer and the amount of total fluid that is produced from the well in the center of the water injectors.

The recovery of oil through water displacement can be considered as (1) displacement, i.e., the fraction of oil displaced from each pore by the physical contact between the displacing water and oil, and (2) conformance, i.e., the fraction of the total volume (area × thickness) that is contacted by the injected water.

Displacement efficiency is the fraction of the oil displaced from each pore, which is contacted by the encroaching water. Areal sweep efficiency is the fraction of oil displaced from the area physically contacted by the encroaching water. The vertical sweep efficiency is the fraction of oil displaced from the vertical cross-sectional coverage by the encroaching water. Hence, the recovery efficiency is the product of displacement efficiency and sweep efficiency (areal efficiency × vertical efficiency).

Waterflooding remains a predominantly secondary recovery process. The principal reason for this is probably that reservoir formation water is ordinarily not available in volume during the early years of an oil field and pressure maintenance water from outside the field may be too expensive. When a young field produces considerable water, it may be injected back into the reservoir primarily for the purpose of nuisance abatement, but reservoir pressure maintenance is a valuable by-product.

Nevertheless, some passages in the formation are larger than others, and the water tends to flow freely through these, bypassing smaller passages where the oil remains. A partial solution to this problem is possible by *miscible fluid flooding*. Liquid butane and propane are pumped into the ground under considerable pressure, dissolving the oil and carrying it out of the smaller passages; additional pressure is obtained by using natural gas.

5.4 Enhanced Oil Recovery

Traditional primary and secondary recovery methods typically recover only one third of the original oil in place. It is at some point before secondary recovery ceases to remain feasible that enhanced oil recovery methods must be applied if further oil is to be recovered.

Enhanced oil recovery (tertiary oil recovery) is the incremental ultimate oil that can be recovered from a petroleum reservoir over oil that can be obtained by primary and secondary recovery methods (Figure 5–8) (Lake, 1989; Arnarnath, 1999). Enhanced oil recovery is often synonymous to some extent with *improved oil recovery* (IOR) as well as *advanced oil recovery* (AOR), although these terms also apply to primary and secondary methods.

Enhanced oil recovery methods have focused on recovering the remaining oil from a reservoir that has been depleted of energy during the application of primary and secondary recovery methods. Much of the easy-to-produce oil has already been recovered from many oil fields, and producers have attempted several tertiary techniques that offer prospects for ultimately producing 30% to 60% (or more) of the reservoir's original oil in place.

Thermal recovery methods include cyclic steam injection, steamflooding (Figure 5–9), use of detergents or surfactants (Figure 5–10), and in situ combustion (Prats, 1986). The steam processes are the most advanced of all enhanced oil recovery methods in terms of field experience, and thus they have the least uncertainty in estimating performance, provided that a good reservoir description is available. Steam processes are most often applied in reservoirs containing viscous oils and tars, usually in place of (rather than following) secondary or primary methods. Commercial application of steam processes has been underway since the early 1960s. In situ combustion has been field tested under a wide variety of reservoir conditions, but few projects have proven economical and advanced to commercial scale.

For tax purposes, the Internal Revenue Service of the United States has listed the projects that qualify as enhanced oil recovery projects (CFR 1.43-2); these thermal recovery methods, gas flood recovery methods, and chemical flood recovery methods are described in the following sections.

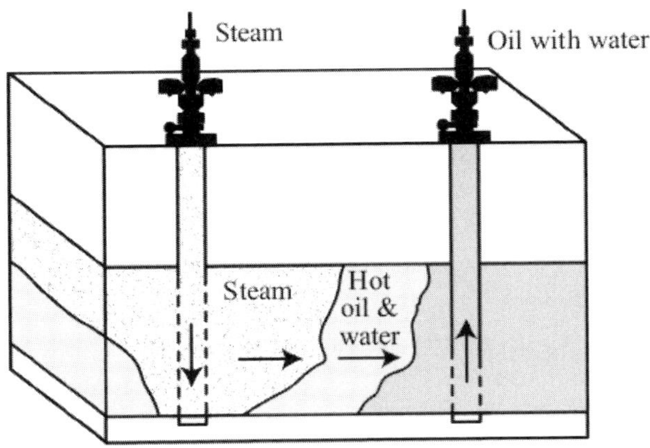

Figure 5–9 *Steamflooding.*

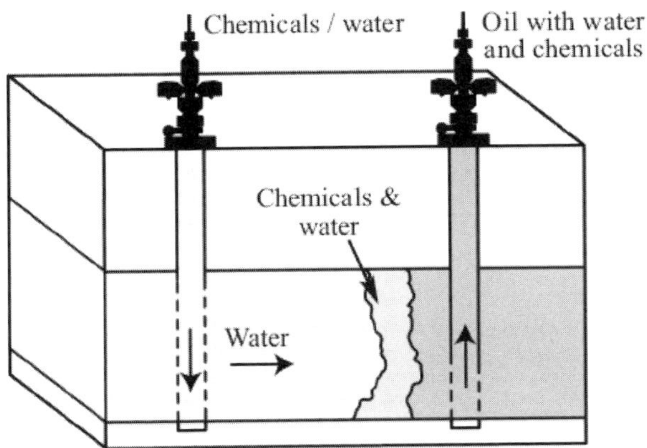

Figure 5–10 *Use of Detergents.*

5.4.1 Thermal Recovery Methods

Thermal enhanced oil recovery processes add heat to the reservoir to reduce oil viscosity and/or to vaporize the oil. In both instances, the oil is made more mobile so that it can be more effectively driven to producing wells. In addition to adding heat, these processes provide a driving force (pressure) to move oil to producing wells.

Steam injection is the most common form of thermally enhanced oil recovery and is used extensively to increase oil production. In situ combustion is another form; instead of using steam to reduce the crude oil viscosity, some of the oil is burned to heat the surrounding oil.

Steam drive injection is the continuous injection of steam into one set of wells (injection wells) or other injection source to effect oil displacement toward and production from a second set of wells (production wells). *Cyclic steam injection* is the alternating injection of steam and production of oil with condensed steam from the same well or wells. *In situ combustion* is the combustion of oil or fuel in the reservoir sustained by injection of air, oxygen-enriched air, oxygen, or supplemental fuel supplied from the surface to displace unburned oil toward producing wells. This process may include the concurrent, alternating, or subsequent injection of water.

Thermal methods for oil recovery have found most use when the oil in the reservoir has a high viscosity. For example, heavy oil is usually highly viscous (hence the use of the adjective *heavy*), with a viscosity ranging from approximately 100 centipoises to several million centipoises at the reservoir conditions. In addition, oil viscosity is also a function of temperature and API gravity (Speight, 2000 and references cited therein). For heavy crude oil samples with API gravity ranging from 4 to 21°API (1.04 to 0.928 kg/m^3),

$$\log \log (\mu\sigma + \alpha) = \mathrm{A} - \mathrm{B} \log (T + 460)$$

where $\mu\sigma$ is oil viscosity in cP, T is temperature in degrees Fahrenheit, A and B are constants, and α is an empirical factor used to achieve a straight-line correlation at low viscosity. This equation is usually used to correlate kinematic viscosity in centistokes, in which case an α of 0.6 to 0.8 is suggested (dynamic viscosity in cP equals kinematic viscosity in cSt times density in g/ml).

An alternative equation for correlating viscosity data is

$$\mu = \mathrm{a}e^{\mathrm{b}/T^*}$$

where a and b are constants, and T^* is the absolute temperature.

Steam drive injection (steam injection) has been commercially applied since the early 1960s. The process occurs in two steps: (1) steam stimulation of production wells, that is, direct steam stimulation, and (2) steam drive by steam injection to increase production from other wells (indirect steam stimulation).

When there is some natural reservoir energy, steam stimulation normally precedes steam drive. In steam stimulation, heat is applied to the reservoir by the injection of high-quality steam into the produce well. This cyclic process, also called *huff and puff* or *steam soak*, uses the same well for both injection and production. The period of steam injection is followed by production of reduced viscosity oil and condensed steam (water). One mechanism that aids production of the oil is the flashing of hot water (originally condensed from steam injected under high pressure) back to steam as pressure is lowered when a well is put back on production.

When natural reservoir drive energy is depleted and productivity declines, most cyclic steam injection projects are converted to steam drives. In some projects, producing wells are periodically steam stimulated to maintain high production rates. Normally, stream drive projects are developed on relatively close well spacing to achieve thermal communication between adjacent injection and production wells. To date, steam methods have been applied almost exclusively in relatively thick reservoirs containing viscous crude oil.

Cyclic steam injection is the alternating injection of steam and production of oil with condensed steam from the same well or wells. Steam generated at the surface is injected in a well and the same well is subsequently put back on production.

A cyclic steam injection process includes three stages. The first stage is injection, during which a measured amount of steam is introduced into the reservoir. The second stage is *the soak period*, which requires that the well be shut in for a period of time (usually several days) to allow uniform heat distribution to reduce the viscosity of the oil (alternatively, to raise the reservoir temperature above the pour point of the oil). Finally, the third stage is production of the now-mobile oil through the same well. The cycle is repeated until the flow of oil diminishes to a point of no returns.

The relatively small portion of the oil that remains after these displacement mechanisms have acted becomes the fuel for the in situ combustion process. Production is obtained from wells offsetting the injection locations. In some applications, the efficiency of the total in situ combustion operation can be improved by alternating water and air injection. The injected water tends to improve the utilization of heat by transferring heat from the rock behind the combustion zone to the rock immediately ahead of the combustion zone.

The performance of in situ combustion is predominantly determined by the four following factors:

1. The quantity of oil that initially resides in the rock to be burned
2. The quantity of air required to burn the portion of the oil that fuels the process
3. The distance to which vigorous combustion can be sustained against heat losses
4. The mobility of the air or combustion product gases

In many field projects, the high gas mobility has limited recovery through its adverse effect on the sweep efficiency of the burning front. Because of the density contrast between air and reservoir liquids, the burning front tends to override the reservoir liquids. To date, combustion has been most effective for the recovery of viscous oils in moderately thick reservoirs in which reservoir dip and continuity provide effective gravity drainage or operational factors permit close well spacing.

Using combustion to stimulate oil production is regarded as attractive for deep reservoirs (Terwilliger, 1975), and in contrast to steam injection, it usually involves no loss of heat. The duration of the combustion may be short (<30 days) or more prolonged (approximately 90 days), depending upon requirements. In addition, backflow of the oil through the hot zone must be prevented or coking occurs.

Both forward and reverse combustion methods have been used with some degree of success when applied to tar sand deposits. The forward combustion process has been applied to the Orinoco deposits (Terwilliger et al., 1975) and in the Kentucky sands (Terwilliger, 1975). The reverse combustion process has been applied to the Orinoco deposits (Burger, 1978) and the Athabasca (Mungen and Nicholls, 1975). In tests such as these, it is essential to control the airflow and

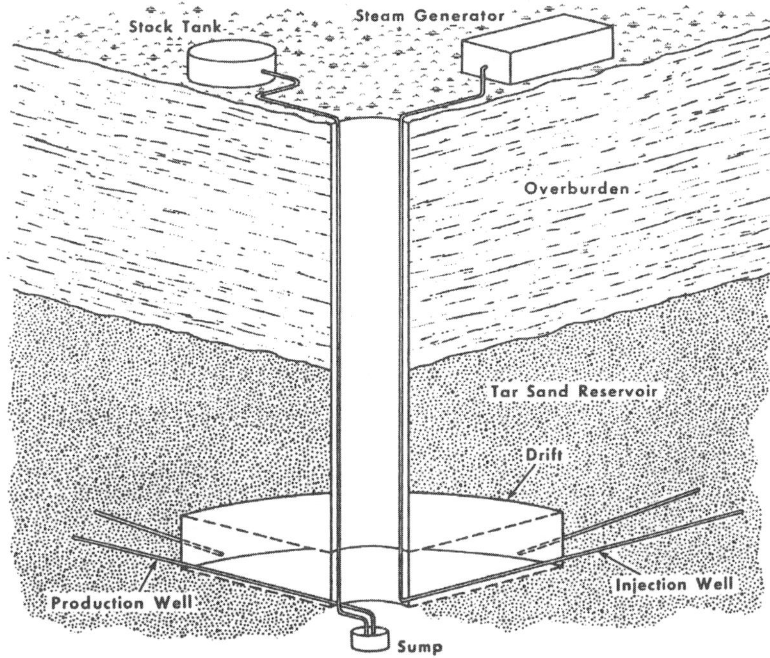

Figure 5–11 *Modified in-situ extraction.*

to mitigate the potential for spontaneous ignition (Burger, 1978). A modified combustion approach has been applied to the Athabasca deposit (Mungen and Nicholls, 1975). The technique involved a heat-up phase and a production (or blowdown) phase followed by a displacement phase using a fireflood-waterflood (COFCAW) process.

In *modified in-situ extraction* processes, a combination of in situ and mining techniques is used to access the reservoir when the heavy oil proves too difficult to move to the production well. Such processes are amenable to recovery of bitumen from tar sand deposits (Figure 5–11). A portion of the reservoir rock must be removed to enable application of the in situ extraction technology. The most common method is to enter the reservoir through a large-diameter vertical shaft, excavate horizontal drifts from the bottom of the shaft, and drill injection and production wells horizontally from the drifts. Thermal extraction processes are then applied through the wells. When the horizontal wells are drilled at or near the base of the tar sand reservoir, the injected heat rises from the injection wells through the reservoir, and drainage of produced fluids to the production wells is assisted by gravity.

Technologies that upgrade value, drive down costs, and reduce environmental impacts will have the greatest effect on increasing the production of heavy oil. There are a large number of technologies that can have an impact, but there is no single technology that can be generally applied, owing to the variability of heavy oil properties as well as reservoir properties. In recent years, there has been a renewed interest in the in situ combustion technologies that have the potential to partially upgrade the oil and produce a product at the surface that is more valuable than the original oil.

Finally, no single enhanced oil recovery technique is the cure-all for oil recovery. Most reservoirs are complex, and the oil-reservoir system must be considered as a whole rather than as individual, but equally complex, entities.

5.4.2 Gas Flood Recovery Methods

Miscible fluid displacement involves the injection of gas (e.g., natural gas, enriched natural gas, a liquefied petroleum slug driven by natural gas, carbon dioxide, nitrogen, or flue gas) or alcohol into the reservoir at pressure levels such that the gas or alcohol and reservoir oil are miscible. *Carbon dioxide augmented waterflooding* involves the injection of carbonated water, or water and carbon dioxide, to increase waterflood efficiency. *Immiscible carbon dioxide displacement* involves the injection of carbon dioxide into an oil reservoir to effect oil displacement under conditions in which miscibility with reservoir oil is not obtained. This process may include the concurrent, alternating, or subsequent injection of water. *Immiscible nonhydrocarbon gas displacement* involves the injection of nonhydrocarbon gas (e.g., nitrogen) into an oil reservoir, under conditions in which miscibility with reservoir oil is not obtained, to obtain a chemical or physical reaction (other than pressure) between the oil and the injected gas or between the oil and other reservoir fluids. This process may include the concurrent, alternating, or subsequent injection of water.

Miscible processes are those in which an injected fluid dissolves in the oil it contacts, forming a single oil-like liquid that can flow through the reservoir more easily than the original crude oil. A variety of such processes have been developed using different fluids that can mix with oil, including alcohols; carbon dioxide; petroleum hydrocarbons such as propane or propane-butane mixtures; and petroleum gases rich in ethane, propane, butane, and pentane. The fluid must be carefully selected for each reservoir and type of crude oil to ensure that the oil and injected fluid will mix. The cost of the injected fluid is usually quite high in all known processes, and therefore either the process must include a supplementary operation to recover expensive injected fluid, or the injected material must be used sparingly. In this process, a slug, which varies from 5% to 50% of the reservoir volume, is pushed through the reservoir by gas, water (brine), or chemically treated brine to contact and displace the mixture of fluid and oil (OTA, 1978).

Miscible processes involve only moderately complex technology compared with other enhanced oil recovery processes. Although many miscible fluids have been field tested, much remains to be determined about the proper formulation of various chemical systems to effect complete volubility and to maintain this volubility in the reservoir as the solvent slug is pushed through it. Because of the high value of hydrocarbons and chemicals derived from hydrocarbons, it is generally felt that such materials would not make desirable injection fluids under current or future economic conditions. For this reason, attention has turned to $C0_2$ as a solvent. Conditions for complete mixing of carbon dioxide with crude oil depend on reservoir temperature and pressure and on the chemical nature and density of the oil.

Miscible fluid displacement (*miscible displacement*) is an oil displacement process in which an alcohol, a refined hydrocarbon, a condensed petroleum gas, carbon dioxide, liquefied natural gas, or even exhaust gas is injected into an oil reservoir, at pressure levels such that the injected gas or alcohol and reservoir oil are miscible; the process may include the concurrent, alternating, or subsequent injection of water (Stalkup, 1983).

Miscible displacement occurs when oil flows to the production well bores from deep within the reservoir, displacing the oil near the well bore or water moving within the aquifer. Oil can also move to the well bore by immiscible displacement when the oil is displaced by water or gas (Caruana and Dawe, 1996; Dawe, 2004). There can also be mass transport across phase boundaries (e.g., gas from liquid to vapor in solution gas processes, liquid dropout from gas in condensate reservoirs, multicontact processes in miscible gas enhanced oil recovery processes, and transfer of chemicals such as in surfactant enhanced oil recovery). Additionally, there can be temperature changes with corresponding physical changes as in thermal enhanced oil recovery and cooling during long-term waterflooding of a reservoir with cold water (Dawe, 2004).

For miscible displacement (e.g. solvent displacing oil in front of an immiscible displacement front or water displacing water behind it), the flow lines generally follow the permeability changes. If the flow lines are not parallel to the heterogeneity, some interesting consequences can occur, and the predictions can become difficult. If immiscible displacements occur, interpretation becomes even more problematic as not only are there permeability and viscous effects but also effects due to capillary forces, wettability, and saturation, which can become dominant. The saturation effects are particularly difficult as they are affected by pore scale wettability, which modify the relative permeability and capillary pressures, which themselves modify the displacement patterns.

When carbon dioxide is used for oil displacement, the carbon dioxide enters the oil-bearing porous media at reservoir temperature, which is usually between 26°C and 120°C (80°F and 250°F). The carbon dioxide is introduced at a pressure that is high enough to enable miscible displacement to be achieved, referred to as the minimum miscibility pressure (MMP). The foremost disadvantage of carbon dioxide as an oil displacement fluid is its low viscosity, 0.03 to 0.10 cp at these reservoir conditions. The brine in a reservoir has a viscosity on the order of 1 cp, while the reservoir oil viscosity varies between 0.1 to 50 cp. The carbon dioxide slug therefore has a much higher mobility (the ratio of fluid permeability in porous media to viscosity) than the fluid it is displacing, which results in an unfavorably low mobility ratio (the ratio of the carbon dioxide mobility to the fluid it is displacing is much less than unity). As a result, the areal sweep efficiency of the flood can be very low as the carbon dioxide *fingers* toward the production wells rather than uniformly displacing the oil ahead of it toward the production wells. Even though the displacement efficiency of the carbon dioxide may be very high for the oil it contacts, these fingers result in the carbon dioxide *bypassing* much of the oil in the reservoir. Consequently, if the carbon dioxide viscosity could be elevated to a level comparable with the oil it is displacing, typically a one to two order of magnitude increase, substantial improvements in sweep efficiency and oil recovery could result.

The procedures for miscible displacement are the same in each case and involve the injection of a slug of solvent that is miscible with the reservoir oil followed by injection of either a liquid or a gas to sweep up any remaining solvent. It must be recognized that the miscible slug of solvent becomes enriched with oil as it passes through the reservoir, and its composition changes, thereby reducing the effective scavenging action. Changes in the composition of the fluid can also lead to deposition of wax (Weingarten and Euchner, 1986; Majeed et al., 1990; Pedersen et al., 1991; Erickson et al., 1993; Pan and Firoozabadi, 1996; Calange, 1997) as well as deposition of asphaltene constituents (Leontaritis et al., 1987 and references cited therein; Leontaritis, 1989; Chung, 1992; Nghiem et al., 1993; Nor-Aziam and Adewumi, 1993; Kamath et al., 1994; Deo et al., 1995; Rassamdana et al., 1999). Therefore, caution is advised.

Microscopic observations of the leading edge of the miscible phase have shown that the displacement takes place at the boundary between the oil and the displacing phase (Koch and Slobod, 1957). The small amount of oil that is bypassed is entrained and dissolved in the rest of the slug of miscible fluids. Mixing and diffusion occur to permit complete recovery of the remaining oil. If a second miscible fluid is used to displace the first, another zone of displacement and mixing follows. The distance between the leading edge of the miscible slug and the bulk of pure solvent increases with the distance traveled, as mixing and reservoir heterogeneity cause the solvent to be dispersed.

Other parameters affecting the miscible displacement process are reservoir length, injection rate, porosity, and permeability of the reservoir matrix; size and mobility ratio of miscible phases; gravitational effects; and chemical reactions. Miscible floods using carbon dioxide, nitrogen, or hydrocarbons as miscible solvents have their greatest potential for enhanced recovery of low-viscosity oils.

During a miscible displacement project, carbon dioxide dynamically develops miscibility as it mixes with the oil in the porous media. This process is conducted at or slightly above the *minimum miscibility pressure* to ensure that the solvent strength of the carbon dioxide is great enough to attain a high degree of solvency for the oil that it contacts. As the reservoir fluids are produced from the reservoir, the carbon dioxide can be readily separated from the oil and brine via pressure reduction. The availability of large volumes of low cost carbon dioxide, along with its low-toxicity, non-flammability and classification as a nonvolatile organic compound combine to make it an attractive fluid for enhanced oil recovery (Enick, 1998). Commercial hydrocarbon-miscible floods have been operated since the 1950s, but carbon dioxide miscible flooding on a large scale is relatively recent and is expected to make the most significant contribution to miscible enhanced recovery in the future.

Carbon dioxide is capable of displacing many crude oils, thus permitting recovery of most of the oil from the reservoir rock that is contacted (*carbon dioxide miscible flooding*). The carbon dioxide is not initially miscible with the oil. However, as the carbon dioxide contacts the in situ crude oil, it extracts some of the hydrocarbon constituents of the oil; carbon dioxide is also dissolved in the oil. Miscibility is achieved at the displacement front when no interfaces exist between the hydrocarbon-enriched carbon dioxide mixture and the carbon dioxide-enriched oil. Thus, by a *dynamic* (*multiple-contact*) process involving interphase mass transfer, miscible displacement overcomes the capillary forces that otherwise trap oil in pores of the rock.

The reservoir operating pressure must be kept at a level high enough to develop and maintain a mixture of carbon dioxide and extracted hydrocarbons that, at reservoir temperature, will be miscible with the crude oil. Impurities in the carbon dioxide stream, such as nitrogen or methane, increase the pressure required for miscibility. Mixing due to reservoir heterogeneity and diffusion tends to locally alter and destroy the miscible composition, which must then be regenerated by additional extraction of hydrocarbons. In field applications, both miscible and near-miscible displacements may actually proceed simultaneously in different parts of the reservoir.

The volume of carbon dioxide injected is specifically chosen for each application and usually ranges from 20% to 40% of the reservoir pore volume. In the later stages of the injection program, carbon dioxide may be driven through the reservoir by water or a lower cost inert gas (Meszaros et al, 1990). To achieve higher sweep efficiency, water and carbon dioxide are often injected in alternate cycles.

In some applications, particularly in carbonate (limestone, dolomite, and chert) reservoirs where it is likely to be used most frequently, carbon dioxide may prematurely break through to producing wells. When this occurs, remedial action using mechanical controls in injection and production wells may be taken to reduce carbon dioxide production. However, substantial carbon dioxide production is considered normal. Generally, this produced carbon dioxide is reinjected, often after processing, to recover valuable light hydrocarbons.

For some reservoirs, miscibility between the carbon dioxide and the oil cannot be achieved; this is dependent upon the oil properties. However, carbon dioxide can still be used to recover additional oil. The carbon dioxide swells crude oils, thus increasing the volume of pore space occupied by the oil and reducing the quantity of oil trapped in the pores. It also reduces the oil viscosity. Both effects improve the mobility of the oil. Carbon dioxide immiscible flooding has been demonstrated in both pilot and commercial projects, but overall it is expected to make a relatively small contribution to enhanced oil recovery.

The solution gas-oil ratio for carbonated crude oil should be measured in the normal way and plotted as gas-oil ratio in volume per volume versus pressure. The greater the solubility of carbon dioxide in the oil, the larger is the increase in the solution gas-oil ratio. In fact, the increase in the gas-oil ratio usually parallels the increase in the oil formation volume factor due to swelling. It should be noted that the gas in any gas-oil ratio experiment is not carbon dioxide but contains hydrocarbons that have vaporized from the liquid phase. Consequently, whether the gas-oil ratio is measured in a pressure-volume-temperature cell or from a slim tube experiment, compositional analysis must be carried out to obtain the composition of the gas as well as that of the equilibrium liquid phase. If actual measured values are not available, the correlation developed for crude oil containing dissolved gases can be used but give only approximate values at best. Since the density of pure gases is a function of pressure and temperature, for crude oil saturated with gases, the density in the mixing zone must be specified as a function of pressure and mixing zone composition.

Hydrocarbon gases and condensates have been used for over 100 commercial and pilot miscible floods. Depending upon the composition of the injected stream and the reservoir crude oil, the mechanism for achieving miscibility with reservoir oil can be similar to that obtained with carbon dioxide (dynamic or multiple-contact miscibility), or the miscible solvent and in situ oil may be miscible initially (first-contact miscibility). Except in special circumstances, these light hydrocarbons are generally too valuable to be used commercially.

Nitrogen and flue gases have also been used for commercial miscible floods. Minimum miscibility pressures for these gases are usually higher than for carbon dioxide, but in high-pressure, high-temperature reservoirs where miscibility can be achieved, these gases may be a cost-effective alternative to carbon dioxide.

5.4.3 Chemical Flood Recovery Methods

Three enhanced oil recovery processes involve the use of chemicals such as caustic, surfactant/polymer, and polymer (OTA, 1978).

Surfactant flooding is a multiple-slug process involving the addition of surface-active chemicals to water (Reed and Healy, 1977). These chemicals reduce the capillary forces that trap the oil in the pores of the rock. The surfactant slug displaces the majority of the oil from the reservoir volume contacted, forming a flowing oil-water bank that is propagated ahead of the surfactant slug. The principal factors that influence the surfactant slug design are interfacial properties, mobility of the slug in relation to the mobility of the oil-water bank, the persistence of acceptable slug properties and slug integrity in the reservoir, and cost.

A slug of water containing polymer in solution follows the surfactant slug. The polymer solution is injected to preserve the integrity of the more costly surfactant slug and to improve the sweep efficiency. Both these goals are achieved by adjusting the polymer solution viscosity in relation to the viscosity of the surfactant slug to obtain a favorable mobility ratio. The polymer solution is then followed by injection of drive water, which continues until the project is completed.

Each reservoir has unique fluid and rock properties, and specific chemical systems must be designed for each individual application. The chemicals used, their concentrations in the slugs, and the slug sizes depend upon the specific properties of the fluids and the rocks involved and upon economic considerations.

Surfactant flooding is complex and requires detailed laboratory testing to support field project design. As demonstrated by field tests, it has excellent potential for improving the recovery of low-viscosity to moderate-viscosity oil. Surfactant flooding is also expensive and has been used in few large-scale projects.

Surfactant-polymer flooding, also known as *microemulsion flooding,* involves injection of a surfactant system (e.g., a surfactant, hydrocarbon, cosurfactant, electrolyte, and water) to enhance the displacement of oil toward producing wells.

The terms *microemulsion* and *micellar solution* are used to describe concentrated, surfactant-stabilized dispersions of water and hydrocarbons that are used to enhance oil recovery. At concentrations above a certain critical value, the surfactant molecules in solution form aggregates called micelles. These micelles are capable of solubilizing fluids in their cores and are called swollen micelles. Spherical micelles have size ranges from 10^{-6} to 10^{-4} mm. The micellar solution or microemulsion is homogeneous, transparent or translucent, and stable to phase separation. The term *soluble oil* is often used to describe an oil external system having little or no dispersed water.

Although used to describe the process, neither microemulsion or micellar solution accurately describes all compositions that are used in emulsion flooding. In fact, many systems used do not have an identifiable external or continuous phase, and these terms do not always apply.

In microemulsion flooding, a stable solution of oil, water, and one or more surfactants along with electrolytes of salts is injected into the formation and is displaced by a mobility buffer solution (Reed and Healy, 1977; Dreher and Gogarty, 1979). Injecting water in turn displaces the mobility buffer. Depending on the reservoir environment, a pre-flood may or may not be used (Venuto, 1989). The microemulsion is the key to the process. Oil and water are displaced ahead of the microemulsion slug, and a stabilized oil-water bank develops. The displacement mechanism is the same under secondary and tertiary recovery conditions. In the secondary case, water is the primary produced fluid until the oil bank reaches the well.

Two approaches have developed to enhance oil recovery in microemulsion flooding (Gogarty, 1976). In the first process, a relatively low-concentration surfactant microemulsion is injected at large pore volumes of 15% to 60% to reduce the interfacial tension between the water and oil, thereby increasing oil recovery. In the second process, a relatively small pore volume, from 3% to 20%, of a high-concentration surfactant microemulsion is injected. With the high concentration of surfactant in the microemulsion, the micelles solubilize the oil and water in the displacing microemulsion. Consequently, the high-concentration system may initially displace the oil in a miscible-like manner. As the high-concentration slug moves through the reservoir, it is diluted by the formation fluids, and the process ultimately or gradually reverts to a low-concentration flood. However, this initial displacement forms an oil bank, which is very important in establishing displacement efficiency. Low-concentration systems typically contain 2% to 4% surfactant w/w, whereas high-concentration systems contain 8% to 12%.

Mobility control is important to the success of the process. The mobility of the microemulsion can be matched to that of the stabilized oil-water bank by controlling the microemulsion viscosity. The mobility buffer following the microemulsion slug prevents rapid slug deterioration from the rear and thus minimizes the slug size required for efficient oil displacement. Water external emulsions and aqueous solutions of high-molecular-weight polymers have been used as mobility buffers.

Microemulsion flooding can be applied over a wide range of reservoir conditions. Generally, wherever a waterflood has been successful, microemulsion flooding may also be applicable. In cases in which waterflooding was a failure because of poor mobility relationships, microemulsion flooding might be technically successful because of the required mobility control. Of course, if waterflooding was a failure because of certain reservoir conditions, such as fracturing or very high permeability streaks, microemulsion flooding will most likely also fail.

In microemulsion flooding, the slug must be designed for specific reservoir conditions of temperature, resident water salinity, and crude oil type. If the temperature is very high, a fluid-handling problem may result in the field because of the increased vapor pressure of the hydrocarbons in the microemulsion.

In analyzing the applicability of microemulsion flooding to a given reservoir, the need for a thorough understanding of the reservoir and fluid characteristics cannot be overemphasized. As mentioned, such characteristics as the nature of the oil and water content; relative permeability; mobility ratios; formation fractures; and variations in permeability, porosity, formation continuity, and rock mineralogy can have a dramatic effect on the success or failure of the process.

Polymer augmented waterflooding (sometimes called *polymer flooding*) involves the injection of polymeric additives with water to improve the areal and vertical sweep efficiencies of the reservoir by increasing the viscosity and decreasing the mobility of the water injected. Polymer augmented waterflooding does not include the injection of polymers for the purpose of modifying the injection profile of the well bore or the relative permeability of various layers of the reservoir, rather than modifying the water-oil mobility ratio.

Conventional waterflooding can often be improved by the addition of polymers to injection water to improve the mobility ratio between the injected and in-place fluids (Sorbie, 1991). The polymer solution affects the relative flow rates of oil and water and sweeps a larger fraction of the reservoir than water alone, thus contacting more of the oil and moving it to production wells. Polymers currently in use are produced both synthetically (polyacrylamides) and biologically (polysaccharides). The polymers may also be cross-linked in situ to form highly viscous fluids that will divert the subsequently injected water into different reservoir strata.

Polymer flooding has its greatest utility in heterogeneous reservoirs and those that contain moderately viscous oils. Oil reservoirs with adverse waterflood mobility ratios have a potential for increased oil recovery through better horizontal sweep efficiency. Heterogeneous reservoirs may respond favorably as a result of improved vertical sweep efficiency. Because the microscopic displacement efficiency is not affected, the increase in recovery over waterflood will likely be modest and limited to the extent that sweep efficiency is improved, but the incremental cost is also moderate. Currently, polymer flooding is being used in a significant number of commercial field projects. The process may be used to recover oils of higher viscosity than those for which a surfactant flood might be considered.

Polymer solutions must be stable for a prolonged period at reservoir conditions. Mechanical, chemical, thermal, and microbial effects can degrade polymers. However, degradation can be minimized or even prevented by using specific equipment or methods.

Stability problems may occur as a result of oxygen contamination of the polymer solutions. Such contamination can lower the screen factor of polyacrylamide solutions by as much as 30%. In field operations, the loss of mobility reduction due to oxygen may be more serious since control of the reservoir fluid composition can be difficult. Sodium hydrosulfite in low concentrations is an effective oxygen collector for polyacrylamide solutions. However, sodium hydrosulfite tends to catalyze polymer deterioration when free oxygen and decomposed polymers are present. Therefore, the proper use of sodium hydrosulfite is imperative to avoid severe polymer degradation. In addition, caution is necessary to prevent oxygen from reentering the system once sodium hydrosulfite has been added to the makeup water.

Alkaline flooding (or caustic flooding) is the injection of water that has been made chemically basic by the addition of alkali metal hydroxides, silicates, or other chemicals.

Alkaline flooding involves the use of aqueous solutions of certain chemicals such as sodium hydroxide, sodium silicate, and sodium carbonate that are strongly alkaline. These solutions will react with constituents present in some crude oils or present at the rock/crude oil interface to form detergentlike materials, which reduce the ability of the formation to retain the oil. These chemicals enhance oil recovery by one or more of the following mechanisms: interfacial tension reduction, spontaneous emulsification, or wettability alteration. These mechanisms rely on the in situ formation of surfactants during the neutralization of petroleum acids in the crude oil by the alkaline chemicals in the displacing fluids.

Although emulsification in alkaline flooding processes decreases injection fluid mobility to a certain degree, emulsification alone may not provide adequate sweep efficiency. Sometimes polymer is included as an ancillary mobility control chemical in an alkaline waterflood to augment any mobility ratio improvements due to alkaline-generated emulsions.

Other variations on this theme include the use of steam and the means of reducing interfacial tension by the use of various solvents (Ali, 1974; Ali and Abad, 1975). The solvent approach has had some success when applied to bitumen recovery from mined tar sand, but when applied to nonmined material, losses of solvent and dissolved bitumen are always an issue. However, this approach should not be rejected out of hand since a novel concept may arise that guarantees minimal (acceptable) losses of bitumen and solvent

Certain reservoir types, such as those with very viscous crude oils and some low-permeability carbonate (limestone, dolomite, or chert) reservoirs, respond poorly to conventional secondary recovery techniques. The viscosity (or the API gravity) of petroleum is an important factor that must be taken into account when heavy oil is recovered from a reservoir.

In these reservoirs, it is desirable to initiate enhanced oil recovery operations as early as possible. This may mean considerably abbreviating conventional secondary recovery operations or bypassing them altogether

A significant amount of laboratory research and field testing has been devoted to developing enhanced oil recovery methods as well as defining the requirements for a successful recovery and the limitations of the various methods. The intent of enhanced oil recovery is to increase the effectiveness of oil removal from pores of the rock (*displacement efficiency*) and to increase the volume of rock contacted by injected fluids (*sweep efficiency*) (Schumacher, 1980).

To understand the phenomenon of enhanced oil recovery, it is helpful to understand the condition in the reservoir after other recovery operations have been exhausted. The oil remaining after conventional recovery operations is retained in the pore space of reservoir rock at a lower concentration than originally existed. In portions of the reservoir that have been contacted or swept by the injection fluid, the residual oil remains as droplets (or ganglia) trapped in either individual pores or clusters of pores. It may also remain as films partly coating the pore walls. Entrapment of this residual oil is predominantly due to capillary and surface forces and to pore geometry.

In the pores of those volumes of reservoir rock that were not well swept by displacing fluids, the oil continues to exist at higher concentrations and may exist as a continuous phase. This macroscopic bypassing of the oil occurs because of reservoir heterogeneity; the placement of wells; and the effects of viscous, gravity, and capillary forces, which act simultaneously in the reservoir. The resultant effect depends upon conditions at individual locations.

The higher the mobility of the displacing fluid relative to that of the oil (i.e., the higher the *mobility ratio*), the greater the propensity for the displacing fluid to bypass oil. Because of fluid density differences, gravity forces cause vertical segregation of the fluids in the reservoir so that water tends to under-run, and gas to over-ride, the oil-containing rock. These mechanisms can be controlled or utilized only to a limited extent in primary and secondary recovery operations.

For example, during waterflooding, the capillary forces that cause the displacement of oil by water can also result in the trapping of residual oil. Particularly important is the faster movement of water through the smaller pore because (1) its smaller diameter increases the capillary force and (2) the oil volume displaced by water is far less. Water moving more rapidly through the small pore will reach a common outlet before all the oil is displaced from the upper large pore. A net capillary force then is exerted on the downstream end of the large pore; this can be likened to closing a back door. Further displacement of oil ceases, trapping oil between the two interfaces. Most of the emphasis in developing chemically enhanced methods has been toward recovering such residual oil as might remain after waterflooding.

Enhanced oil recovery processes use *thermal, chemical,* or *fluid phase behavior* effects to reduce or eliminate the capillary forces that trap oil within pores, to thin the oil or otherwise improve its mobility, or to alter the mobility of the displacing fluids. In some cases, the effects of gravity forces (which ordinarily cause vertical segregation of fluids of different densities) can be minimized or even used to advantage. The various processes differ considerably in complexity, the physical mechanisms responsible for oil recovery, and the amount of experience that has been derived from field application. The degree to which the enhanced oil recovery methods are applicable in the future will depend on development of improved process technology. It will also depend on improved understanding of fluid chemistry, phase behavior, and physical properties and on the accuracy of geology and reservoir engineering in characterizing the physical nature of individual reservoirs (Borchardt and Yen, 1989).

Polymer flooding (*Polymer augmented waterflooding*) is waterflooding in which organic polymers are injected with the water to improve horizontal and vertical sweep efficiency. Alkaline flooding has been used only in those reservoirs containing specific types of high-acid-number crude oils.

5.5 References

Ali, S.M. Farouq. 1974. Oil Sands Fuel of the Future, ed. L.V. Hills, 199. Canadian Society of Petroleum Geologists, Calgary, Alberta.

Ali, S.M. Farouq, and Abad, B. 1975. Proceedings of the Twenty-sixth Annual Meeting, Petroleum Society, Canadian Institute of Mining. Banff, Alberta, Canada.

Arnarnath, A. 1999. Enhanced Oil Recovery Scoping Study. Report No. TR-113836. Electric Power Research Institute, Palo Alto, California.

Borchardt, J.K., and Yen, T.F. 1989. Oil Field Chemistry. Symposium Series No. 396. American Chemical Society, Washington, DC.

Burger, J. 1978. Developments in Petroleum Science, No. 7, Bitumens, Asphalts and Tar Sands, ed. G.V. Chilingarian and T.F. Yen. Elsevier, New York.

Calange, S., Ruffier-Meray, V., and Behar, E. 1997. Onset of Crystallization Temperature and Deposit Amount for Waxy Crudes: Experimental Determinations and Thermodynamic Modeling. Paper No. SPE 37239. Proceedings. Annual Technical Conference and Exhibition. Houston, Texas, February 18–21.

Caruana, A., and Dawe, R.A. 1996. Effect of Heterogeneities on Miscible and Immiscible Flow Processes in Porous Media. Trends in Chemical Engineering. 3:185–203.

CFR 1.43-2. 2004. Internal Revenue Service, U.S. Department of the Treasury, Washington DC.

Chakma, A., Islam, M.R., and Berruti, F., ed. 1991. Enhanced Oil Recovery. AIChE Symposium Series. 87(280). American Institute of Chemical Engineers, New York.

Chung, T-H. 1992. Thermodynamic Modeling for Organic Solids Precipitation. Paper No. SPE 24851. Proceedings. 67th Annual Technical Conference. Washington, DC, October 4–7.

Craft, B.C., and Hawkins, M.F. 1959. Applied Petroleum Reservoir Engineering. Prentice-Hall, Englewood Cliffs, New Jersey.

Dawe, R.A. 2004. Miscible Displacement In Heterogeneous Porous Media. Proceedings. Sixth Caribbean Congress Of Fluid Dynamics. Uwi, January 22–23.

Deo, M.D., Miharia, A., and Kumar, R. 1995. Solids Precipitation in Reservoirs Due to Non-isothermal Injections. Paper No. 28967. Proceedings. San Antonio, Texas. February 14–17.

Dreher, K.D., and Gogarty, W.B. 1979. Journal of Rheology. 23(2):209.

Drew, L.J. 1997. Undiscovered Mineral and Petroleum Deposits: Assessment & Controversy. Plenum Publishing Corporation, New York.

Enick, R.M. 1998. A Literature Review of Attempts to Increase the Viscosity of Dense Carbon Dioxide. Contract DE-AP26-97FT25356. U.S. Department of Energy, Federal Energy Technology Center, Morgantown, West Virginia.

Erickson, D.D., Nielsen, V.G., and Brown, T.S. 1993. Thermodynamic Measurement and Prediction of Paraffin Precipitation in Crude Oil. Paper

No. SPE 26604. Proceedings. 68th Annual Technical Conference and Exhibition. Houston, Texas, October 3–6.

Frick, T.C. 1962. Petroleum Production Handbook. Volume II. McGraw-Hill, New York.

Gogarty, W.B. 1976. Journal of Petroleum Technology. 93.

Jayasekera, A.J., and Goodyear, S.G. 1999. The Development of Heavy Oil Fields in the UK Continental Shelf: Past, Present, and Future. SPE 54623. Society of Petroleum Engineers International Thermal Operations and Heavy Oil Symposium, Bakersfield California, March.

Kamath, V.A., Kakade, M.G., and Sharma, G.D. 1994. An Improved Molecular Thermodynamic Model for Asphaltene Equilibria. In Asphaltene Particles in Fossil Fuel Exploration, Recovery, Refining, and Production Processes, ed. M.K. Sharma and T.F. Yen. Plenum Press, New York.

Koch, H.A., and Slobod, R.L. 1957. Transactions of the American Institute of Mechanical Engineers. 210:40.

Lake, L.W. 1989. Enhanced Oil Recovery. Prentice-Hall Inc., Englewood Cliffs, New Jersey.

Lake, L.W., and Walsh, M.P. 2004. Primary Hydrocarbon Recovery. Elsevier, Amsterdam, The Netherlands.

Leontaritis, K.J. 1989. Asphaltene Deposition: A Comprehensive Description of the Problem, Manifestations, and Modeling Approaches. Paper No. SPE 18892. Proceedings. Symposium on Production Operations. Oklahoma City, Oklahoma, March 13–14.

Leontaritis, K.J., Mansoori, G.A., and Mansoori, U. 1987. Asphaltene Flocculation during Oil Production and Processing: A Thermodynamic Colloidal Model. Paper No. SPE 16258. Proceedings. International Symposium on Oilfield Chemistry. San Antonio, Texas, February 4–6.

Majeed, A., Bringedal, B., and Overå, S. 1990. Oil & Gas Journal. June 18:63–69.

Meszaros, G., Chakma, A., Jha, K.N., Islam, M.R. 1990 Scaled Model Studies and Numerical Simulation of Inert Gas Injection with Horizontal Wells. Proceedings. SPE Annual Technical Conference, EOR Volume, 585–598. New Orleans, Louisiana, September 23–26.

Mungen, R., and Nichols, J.H. 1975. Proceedings. 9th World Petroleum Congress. Elsevier, Amsterdam, The Netherlands. Volume 5:29.

Nghiem, L.X., Hassam, M.S., Nutakki, R., and George, A.E.D. 1993. Efficient Modeling of Asphaltene Precipitation. Paper No. SPE 26642. Proceedings. 68th Annual Technical Conference. Houston, Texas, October 4–7.

Nor-Aziam, N., and Adewumi, M.A. 1993. Development of Asphaltene Phase Equilibria Predictive Model. Paper No. SPE 26905. Proceedings. Eastern

Regional Conference and Exhibition. Pittsburgh, Pennsylvania, November 2–4.

Office of Technology Assessment. 1978. Enhanced Oil Recovery Potential in the United States. NTIS order #PB-276594. Office of Technology Assessment, Washington, DC.

Pan, H., and Firoozabadi, A. 1996. Pressure and Composition Effects on Wax Precipitation: Experimental Data and Model Results. Paper No. SPE 36740. Proceedings. Annual Technical Conference and Exhibition. Denver, Colorado, October 6–9.

Pedersen, K.S., Skovborg, P., and Rønningsen, H.P. 1991. Energy & Fuels. 5:924–932.

Prats, M. 1986. Thermal Recovery. Volume 7. Society of Petroleum Engineers, New York.

Rassamdana, H., Farhani, M., Dabir, B., Mozaffarian, M., and Sahimi, M.1999. Energy & Fuels. 13:176–187.

Reed, R.L., and Healy, R.N. 1977. Improved Oil Recovery by Surfactant and Polymer Flooding. Ed. D.O. Shah and R.S. Schechter. Academic Press, New York.

Schumacher, M.M. 1980. Enhanced Recovery of Residual and Heavy Oils. Noyes Data Corp., Park Ridge, New Jersey.

Sorbie, K.S. 1991. Polymer-Improved Oil Recovery. Springer, New York.

Speight, J.G. 2000. Desulfurization of Heavy Oils and Residua. 2nd Edition. Marcel Dekker Inc., New York.

Stalkup, F.I., 1983. Miscible Displacement. SPE Monograph No. 8. Society of Petroleum Engineers, New York.

Taber, J.J., and Martin, F.D. 1983. Technical Screening Guides for the Enhanced Recovery of Oil. Paper No. SPE 12069. Proceedings. 58th SPE Annual Technical Conference and Exhibition. San Francisco, California, October 5–8.

Terwilliger, P.L. 1975. Paper No. 5568. Proceedings. 50th Annual Fall Meeting. Society of Petroleum Engineers, American Institute of Mechanical Engineers, Dallas Texas, September.

Terwilliger, P.L., Clay, R.R., Wilson, L.A., and Gonzalez-Gerth, E. 1975. Journal of Petroleum Technology. 27:9.

U.S. Department of Energy. 1996. Maintaining Oil Production from Marginal Fields: A Review of the Department of Energy's Reservoir Class Program. Panel on the Review of the Oil Recovery Demonstration Program of the Department of Energy, Committee on Earth Resources, Board on Earth Sciences and Resources, Commission on Geosciences, Environment, and

Resources, National Research Council. National Academy Press, Washington, DC.

Venuto, P.B. 1989. Tailoring EOR Processes to Geologic Environments. World Oil. 209(Nov):61–68.

Weingarten, J.S., and Euchner, J.A. 1986. Methods for Predicting Wax Precipitation and Deposition. SPE Paper No. 15654. Proceedings. 61st Annual Technical Conference and Exhibition. New Orleans, Louisiana, October 5–8.

Yang, C., Gu, Y. 2005a A Novel Experimental Technique for Studying Solvent Mass Transfer and Oil Swelling Effect in a Vapor Extraction (VAPEX) Process. Paper No. 2005-099. Proceedings. 56th Annual Technical Meeting, The Canadian International Petroleum Conference. Calgary, Alberta, Canada, June 7–9.

Yang, C., and Gu, Y. 2005b. Effects of Solvent-Heavy Oil Interfacial Tension on Gravity Drainage in the VAPEX Process. Paper No. SPE 97906. Society of Petroleum Engineers International Thermal Operations and Heavy Oil Symposium. Calgary, Alberta, Canada, November 1–3.

CHAPTER 6

NONTHERMAL METHODS OF RECOVERY

The vast heavy oil reserves available in various parts of the world are becoming increasingly important as a secure future energy source (Safinya, 2008). In fact, the heavy oil resource is trillions of barrels, but the cumulative recovery totals to date are on the order of billions of barrels. Whether the potential and promise of heavy oil is realized depends on the evolution of recovery technologies that are appropriate for the wide range of reservoir and oil-phase conditions. Such technologies also need to be comparatively benign from an environmental view.

Heavy oil production has been increasing in recent years and is expected to increase in the future because of expected supply shortfall in conventional oil and an abundance of relatively large and known heavy oil reservoirs. Heavy oil is commercially produced by primary recovery, water injection, and thermal enhanced oil recovery methods. A majority of the recent primary production is in Venezuela's Orinoco belt; long horizontal wells have made these projects possible.

Steam injection is the most widely applied enhanced oil recovery method. A majority of the steam injection projects are in California, Canada, Indonesia, and Venezuela. Recovery can approach 20% for cyclic steaming and over 50% for continuous steam injection (Kumar, 2006). Novel methods such as solvent injection and hybrid methods are being tested for heavy oil recovery, where steam may not be the best option.

However, heavy oil in a reservoir is often very viscous (unless the reservoir temperature is high) and does not flow easily, therefore traditional methods of oil recovery, such as primary and secondary methods, are not usually applicable to heavy oil. If they are applicable, it is often decided to bypass such methods and proceed to enhanced methods of recovery. The route to use is, of course, reservoir specific, and what is applied to one heavy oil reservoir cannot necessarily be applied to another heavy oil reservoir.

The amount of oil that is recoverable is determined by a number of factors, including the permeability of the rocks, the strength of natural drives (the gas present, pressure from adjacent water or gravity), and the viscosity of the oil. When the reservoir rocks are "tight" such as in shale, oil generally cannot flow through, but when they are permeable such as in sandstone, oil flows freely. The flow of oil is often helped by natural pressures surrounding the reservoir rocks, including natural gas that may be dissolved in the oil, natural gas present above the oil, water below the oil, and the strength of gravity. Heavy oils tend to span a large range of viscosity from manageable liquids to poorly mobile materials.

In terms of heavy oil recovery, it is more traditional to apply thermal oil recovery techniques; steam injection is the most popular. Heavy oil recovery efforts include thermal methods (steam floods, cyclic steam stimulation, steam assisted gravity drainage SAGD) as well as nonthermal methods (cold flow with sand production, cyclic solvent process, VAPEX). Significant improvements to the effectiveness of these methods can be achieved by developing a basic understanding of the complex displacement mechanisms and by developing new techniques for in situ characterization of fluid and reservoir characteristics. Development of optimal strategies for recovering these reserves requires the development of state-of-the-art reservoir flow simulator(s) that incorporate techniques for coupling geomechanics and fluid flow as well as for accurately representing thermal, phase equilibria, and mass transfer effects. Improved recovery efficiency can also be achieved by various combinations of thermal and nonthermal processes. In addition, recent advances in drilling and production from unconsolidated sands allow development of heavy oil recovery strategies that were not been possible three decades ago.

The quest to produce heavy oil is a global issue. In many cases, the resource does not resemble the conditions that are thought to be typical for heavy oil viscous oil that is held in relatively permeable, shallow sands and the fields of interest have evolved to include fractured carbonates, offshore settings, and deeper, more geologically heterogeneous heavy oil resources. Such new settings introduce new challenges, which include highly permeable pathways and limits on reservoir access, in addition to large oil-phase viscosity and low reservoir energy.

Waterflooding of heavy oil is often summarily dismissed because of adverse mobility ratios. In fact, waterflooding can be used in some heavy oil reservoirs to maintain pressure during cold production. The Captain Field in the North Sea uses horizontal wells for the heavy oil production and horizontal wells for water injection (Etebar, 1995). The horizontal injectors provide more uniformly distributed pressures and a more efficient line-water drive. However, since the viscosity of water is much lower than the heavy oil, care must be taken to avoid water fingering from the injecting wells to the producing wells.

In cold and/or offshore environments, waterflooding (and perhaps polymer-augmented waterflooding) may present the most attractive recovery option following primary recovery. Steam injection is relatively mature, whereas cost-effective mobility and profile control by use of aqueous-phase surfactants, gels, or advanced well completions remains an open question. In situ combustion achieved by air injection is technically and economically feasible, especially for deeper, thinner, higher-pressured reservoirs. Combustion is difficult to describe and control, however. Nevertheless, it is attractive for in situ upgrading and sulfur removal.

6.1 Primary Recovery (Natural) Methods

Primary recovery (Chapter 5) of heavy oil from a reservoir is not the usual approach to recovery. Primary recovery techniques rely entirely on natural forces within the reservoir, and although they are widely applicable to the recovery of conventional crude oil, they are less applicable to heavy oil. At high reservoir temperature that keeps the heavy oil sufficiently fluid, primary recovery may undoubtedly be applicable. Unfortunately, production mechanisms of the heavy oil solution-gas-drive process are not completely elucidated, so it is difficult to optimize primary recovery.

Given the option, operators will try to produce as much heavy oil as possible by primary recovery methods (*cold production*), but this is dependent upon the fluidity of the oil, which in turn is dependent upon the reservoir temperature. The term *cold production* refers to the use of operating techniques and specialized pumping equipment to aggressively produce heavy oil reservoirs without applying heat (Chugh et al., 2000). This encourages the associated production of large quantities of the unconsolidated uncemented reservoir sand, which in turn results in significantly higher oil production.

Typical recovery factors (percentage of the oil in a reservoir that can be recovered) for cold production range from 1 to 10% of the original oil in place (Curtis et al., 2002). In addition, again depending on the character of the reservoir *and* the properties of the oil in the reservoir, primary production with artificial lift (e.g., injection of a light oil, or diluent, to reduce viscosity) may find use. Many fields produce most efficiently with horizontal production wells. In some cases, exploiting foamy oil behavior and/or encouraging the production of sand along with the oil turns out to be the preferred production strategy. Choosing the optimal cold production strategy requires an understanding of rock and fluid properties (Chen, 2006).

Foaming can complicate the recovery process even more. When heavy oil contains associated gas, foaming can occur at any point in the reservoir, well bore, flowlines, or production equipment at the surface. When gas reaches the bubble point and comes out of suspension as the pressure and temperature of the fluids change, the character of the oil changes from the known to the unknown. However, while foaming can be a problem, it can also work in the operator's favor. Since any foam that forms in the reservoir increases pressure, it can serve as a temporary gas drive, pushing oil toward the well bore. The issue is to understand enough about the fluid properties to know where and when the foaming is likely to occur and how the phenomenon can be used beneficially.

A cold production oil reservoir is a *solution gas drive reservoir* if the major reservoir energy for primary depletion is supplied by the release of gas from the oil and the expansion of the in-place fluids as reservoir pressure declines. The fraction of original oil in place that can be recovered by solution gas drive decreases with increasing oil viscosity. For heavy oil reservoirs, the expected recovery factor by solution gas drive is typically about 5%. A number of heavy oil reservoirs under solution gas drive, however, have obtained anomalous primary performance results: low production gas-oil ratios, high oil production rates, and recovery of unexpectedly large amounts of oil. This unusual behavior has been attributed to (1) the expansion of gas bubbles, which gives the oil a foamy character as the bubbles are trapped by the oil and enhances recovery by solution (ultimate oil recovery with primary techniques can be as high as 20% for some heavy foamy oil reservoirs) and (2) internal erosion in unconsolidated sand reservoirs, which can create a network of high-permeability channels (*wormholes*) and enhance drainage by a factor of 10 or more (Chen, 2006).

On the other hand, there are several very large projects (>100,000 barrels per day) for heavy oil (approximately 12° API) recovery in the Heavy Oil Belt (FAJA) in Venezuela, where the main recovery method to date has been primary recovery. The recovery efficiencies are projected to be on the order of 8 to 15% recovery. The foamy nature of the oil has yielded initial rates of over 1000 barrels per day (this is not a common recovery for this gravity oil). Heavy oil production from this belt is expected to last for 35 years at a production rate of 600,000 barrels per day (Meyer and Attanasi, 2003). There are a few factors and technical advances that allow this heavy oil to be produced (Curtis et al., 2002).

In the case of the FAJA oil, first, the viscosity is low enough with the existing solution gas that the heavy oil can flow at reservoir temperatures. Second, horizontal wells up to 1,500 m long allow the heavy oil to be produced at economic rates while maintaining sufficiently low drawdown pressures to prevent extensive sand production. More complex well geometries are being drilled with several horizontal branches (multilateral wells). Third, the horizontal legs are placed precisely in the target sands using logging-while-drilling (LWD) (LWD) and measurement-while-drilling (MWD) (MWD) equipment, enabling more cost effective placement of the wells. Fourth, in some locations, sand production from the unconsolidated formation is minimized using slotted liners and other sand-control methods. A low drawdown pressure in a long multilateral can also reduce the need for significant sand control.

Primary production in some heavy oil reservoirs is larger than that estimated by conventional calculations (Kovscek, 2002). Conventionally, the main driving force behind primary recovery is pressure depletion through solution gas drive. Solution gas drive is the mechanism whereby the lowering of reservoir pressure through production in an undersaturated reservoir causes the oil to reach the bubble point where gas starts to evolve from solution. The evolved gas does not begin to flow until the critical gas saturation has been reached. Once the critical gas saturation point is reached, there is an increase in the rate of pressure drop due to the production of the gas phase. It has been noted that the oil at the well head of these heavy oil reservoirs resembles the foam, hence the term foamy oil. A key to developing an accurate mechanistic understanding of heavy oil solution gas drive is to delineate bubble growth, interaction, and gas flow experimentally.

The main issue for cold production is the low recovery factor (typically less than 15% of the oil in place). Fields are not being developed with future, secondary processes in mind. For example, wells, cement, and completions are not designed for high temperatures encountered in steam injection and other thermal recovery processes. Horizontal and fishbone wells should be drilled in the optimum location with regard to permeability, porosity, oil composition, distances above water or below gas, and the length of the laterals.

Drilling, measurement-while-drilling, and logging-while-drilling technologies are key enablers for this. In horizontal wells and multilateral wells, being able to monitor, understand, control, and ensure the flow from different sections of the well will improve production and reduce unwanted water and/or natural gas production. In the Orinoco field, natural gas production is an issue because of interference with progressive cavity pumps' ability to lift the heavy oil.

6.2 Secondary Recovery Methods

Even in conventional oil reservoirs, a lot of oil can be left behind after primary production since the natural reservoir pressure has dwindled to the point where it can't force the oil to the surface. When dealing with heavy oil reservoirs, 90% or more of the original oil in place can be left in the reservoir after attempts at primary or cold production (Curtis et al., 2002).

As fluid withdrawal continues from the reservoir, the pressure within the reservoir gradually decreases, and the amount of gas in solution decreases. As a result, the flow rate of fluid into the well bore decreases, and less gas is liberated. The fluid may not reach the surface, so a pump (artificial lift) must be installed in the well bore to continue producing the crude oil. If this is allowed to continue, the flow rate of the crude oil becomes so small, and the cost of lifting the oil to the surface becomes so great, that the well costs more to operate than the revenues that can be gained from selling the crude oil (after discounting the price for operating costs, taxes, insurance, and return on capital). The well's economic limit has then been reached, and the well is abandoned. If economical, as it often is, the remaining oil in the well is extracted using secondary oil recovery methods. However, in order to maintain well production, the flow of oil should not be allowed to decrease to the minimum. Operators usually apply secondary recovery at a point before reservoir energy is depleted.

Secondary oil recovery (Chapter 5) uses various techniques to aid in recovering oil from depleted or low-pressure reservoirs. Pumps, such as horsehead pumps and electrical submersible pumps (ESPs), are used to bring the oil to the surface. Other secondary recovery techniques increase the reservoir's pressure by water injection (waterflooding); natural gas injection (gas flooding), which injects air, carbon dioxide, or some other gas into the reservoir—this is not to be confused with gas lift where gas is injected into the annulus of the well rather than the reservoir. Together, primary and secondary recovery methods generally allow 25% to 35% of the reservoir's oil to be recovered.

6.2.1 Waterflooding

Waterflooding is a form of oil recovery (Chapter 5) wherein the energy required to move the oil from the reservoir rock into a producing well is supplied from the surface by means of water injection and the induced pressure from the presence of additional water. Water injection is used to prevent low pressure in the reservoir. The water replaces the oil that has been taken, keeping the production rate and the pressure the same over the long term. Waterfloods are essentially artificial water drives and, at one time, were considered to be a form of enhanced recovery.

In a completely developed oil field, the wells may be drilled anywhere from 200 to 2,000 feet (60 to 600 meters) from one another, depending on the nature of the reservoir. If water is pumped into alternate wells in such a field, the pressure in the reservoir as a whole can be maintained or even increased. In this way, the rate of production of the crude oil also can be increased. In addition, the water physically displaces the oil, thus increasing the recovery efficiency. In some reservoirs with a high degree of uniformity and little clay content, waterflooding may increase the recovery efficiency to as much as 60% or more of the original oil in place. Waterflooding was first introduced in the Pennsylvania oil fields, more or less accidentally, in the late 19th century, and it has since spread throughout the world.

Waterflood has been conducted successfully in a few high viscosity reservoirs in the past, and several projects are currently ongoing and planned around the world. Incremental recovery of approximately 2 to 20% of the original oil in place has been reported (Kumar 2006). The ratio of oil recovered by waterflood to oil recovered by primary methods varies greatly with depth for heavy oil as it does for conventional oil.

Primary recovery from pure gas reservoirs often exceeds 90%. If one is unfortunate enough to have a natural water drive, recovery seldom exceeds 50%. Clearly, water drives (either natural or artificial) are not advantageous in true gas reservoirs.

Any and every source of bulk water can be (and has been) used for injection. Some aspects to consider when selecting an injection source are described in the following sections.

Produced water is often used as an injection fluid. This reduces the potential of causing formation damage due to incompatible fluids, although the risk of scaling or corrosion in injection flowlines or tubing remains. Also, the produced water, being contaminated with hydrocarbons and solids, must be disposed of in some manner, and disposal to sea or river will require a certain level of clean-up of the water stream first. The processing required to render produced water fit for reinjection may be equally costly. As the volumes of water being produced are never sufficient to replace all the production volumes (oil and gas, in addition to water), additional "make-up" water must be provided. Mixing waters from different sources exacerbates the risk of scaling.

Seawater is obviously the most convenient source for offshore production facilities, and it may be pumped inshore for use in land fields. Where possible, the water intake is placed at sufficient depth to reduce the concentration of algae; however, filtering, deoxygenation and biociding is generally required. *Aquifer water* from water-bearing formations other than the oil reservoir, but in the same structure, has the advantage of purity where available. *River water* will always require filtering and biociding before injection.

Filtering must clean the water and remove any impurities, such as shells and algae. Typical filtration is to 2 µm. The filters are so fine so as not to block the pores of the reservoir. Sand filters are the easiest to use because there is an automatic system with Delta P that cleans the filter with a backwash when the sand filter is dirty. The sand filter has different beds with various sizes of sand granules. The water traverses the first, finest, layer of sand down to the coarsest. To clean the filter, the process is inverted. After the water is filtered, it continues on to fill the deoxygenation tower.

Oxygen must be removed from the water because it promotes corrosion and growth of certain bacteria. Bacterial growth in the reservoir can produce toxic hydrogen sulfide, a source of serious production problems, and can block the pores in the rock.

A *deoxygenation tower* brings the injection water into contact with a dry gas stream (gas is always readily available in the oil field). The filtered water drops into the deoxygenation tower, splashing onto a series of trays, causing dissolved oxygen to be lost to the gas stream. An alternative method, also used as a backup to deoxygenation towers, is to add an oxygen scavenging agent such as sodium bisulfite.

The high-pressure, high-flow *water injection pump*s are placed near to the deoxygenation tower and boosting pumps. They fill the bottom of the reservoir with the filtered water to push the oil towards the wells like a piston. The result of the injection is not quick; it needs time.

Since water viscosity (~1 cp) is much lower than the heavy oil (80 to 100 cp), care must be taken to avoid water fingering from the injecting wells to the producing wells.

6.2.2 Gas Injection

Gas injection (also called *reinjection* or *gas repressurization*) is the reinjection of natural gas into an underground reservoir, typically one already containing both natural gas and crude oil, in order to increase the pressure within the reservoir and thus induce the flow of crude oil or else sequester gas that cannot be exported. After the crude oil has been pumped out, the natural gas is once again recovered. Since many of the wells found around the world contain heavy crude oil, this process increases their production.

Recycling of natural gas or other inert gases causes the pressure to rise in the well. This causes more gas molecules to dissolve in the oil, lowering its viscosity and thereby increasing the well's output. Air is not suitable for repressuring wells because it tends to cause deterioration of the oil; instead, carbon dioxide or natural gas is used to repressure the well. Gas reinjection is also sometimes referred to as repressurization; the latter term used only to imply that the pressure inside the well is being increased to aid recovery.

Injection or reinjection of carbon dioxide also takes place in order to reduce the emission of CO_2 into the atmosphere, a form of carbon sequestration. This has been mooted as a major weapon in the future fight against climate change as it allows mass storage of CO_2 over a geological timescale.

6.2.3 Cold Production

Conventional practices of primary heavy oil production discourage sand production and result in minimized initial unit operating costs. This practice, however, may prevent many wells from achieving their maximum oil production rate and reserve potential.

The basis of cold production is that the oil production and recovery improve when sand production occurs naturally. Field production data indicates that heavy oil flows more efficiently when sand is produced from unconsolidated reservoirs. In the Elk Point and Lindberg reservoirs, sand production from wells occurs regularly (Loughead, 1992; McCaffrey and Bowman, 1991).

Sand production is thought to be a function of (1) the absence of clays and cementation materials, (2) the viscosity of the oil, (3) the producing water cut and gas/oil ratio, and (4) the rate of pressure drawdown (Chugh et al., 2000).

The presence of clay stabilizes the sand grains and reduces sand movement. Higher-viscosity oil increases the frictional drag between the oil and the sand grains, which promotes sand movement. High gas or water production inhibits sand production because gas/water is produced instead of an oil/sand mixture. Increasing the drawdown rate also promotes sand movement because of the increase in the velocity of the fluid into the well bore and hence the increase of frictional drag on the sand grains. It has been reported that gross near-well-bore failure of the formation due to sand production results in excellent productivity.

The produced sand creates a modified well-bore geometry that could have several configurations, including piping tubes (wormholes), dilated zones, sheared zones, or (possibly) cavities. Porosity in the dilated zones may increase, leading to large increases in reservoir permeability. In addition, the flow of sand with the oil has the potential to reduce the frictional drag forces on the oil and result in increased productivity in the porous region. Furthermore, fines migration, which occurs during oil production, can block pore throats and reduce the number of flow paths available for the oil. Producing sand helps to eliminate many of these bottlenecks, and the dilation of the sand also creates larger pore throats that are more difficult to block.

Cold heavy oil production with sand (CHOPS) is now widely used as a production approach in unconsolidated sandstones. The process results in the development of high-permeability channels (*wormholes*) in the adjacent low-cohesive-strength sands, facilitating the flow of oil foam that is caused by solution gas drive.

The key benefits of the process are improved reservoir access, order-of-magnitude higher oil production rates (as compared to primary recovery), and lower production costs. The outstanding technical issues involve sand handling problems, field development strategies, wormhole plugging for water shut-off, low ultimate recovery, and sand disposal. Originally, cold production mechanisms were thought to apply only to vertical wells with high-capacity pumps. It is now believed that these mechanisms may also apply to horizontal wells and lighter (heavy) oils.

Instead of blocking sand ingress by screens or gravel packs, CHOPS encourages sand to enter the well bore by aggressive perforation and swabbing strategies. Vertical or slightly inclined wells (vertical to 45°) are operated with rotary progressive cavity pumps (rather than reciprocating pumps) and old fields are converting to higher-capacity progressive cavity pumps, giving production boosts to old wells. Productivity increases over conventional production; a CHOPS process can produce as much as 12 to 25% of the original oil in place, rather than the 0 to 5% typical of primary production without sand. Finally, because massive sand production creates a large disturbed zone, the reservoir may be positively affected for later implementation of thermal processes.

The CHOPS process increases productivity for the following reasons:

- If the sand can move or is unconsolidated, the basic permeability to fluids is enhanced.
- As more sand is produced, a growing zone of greater permeability is generated, similar to a large-radius well, which gives better production.
- Gas coming out of solution in heavy oil does not generate a continuous gas phase. Instead, bubbles flow with the fluid and do not coalesce but expand down-gradient, generating an "internal" gas drive, referred to as *foamy flow*. This also helps to locally destabilize the sand, sustaining the process.
- Continuous sand production means that asphaltene or fines plugging of the near-well-bore environment potentially do not occur, so there is no possibility of an effect to impair productivity.
- As sand is removed, the overburden weight acts to shear and destabilize the sand, helping to drive sand and oil toward the well bore.

Typically, a well placed on CHOPS production will initially produce a high percentage of sand, greater than 20% by volume of liquids. However, this generally drops after some weeks or months. The huge volumes of sand are disposed of by slurry fracture injection, salt cavern placement, or sand placement in a landfill in an environmentally acceptable manner.

CHOPS is used for thin subsurface oil sands (typically 1 to 7 m thick) in Canada, provided the oil sand is unconsolidated and provided the heavy oil contains sufficient solution gas to power the production process. To have any natural gas in solution, the oil sand must be at least a few hundred meters deep. For example, there are a large number of CHOPS wells located near Lloydminster, Alberta. At the time of writing, CHOPS is the only commercial method for exploiting thin oil sands.

CHOPS wells (by definition) require sand production. Foamy oil production may occur without sand production in other areas, such as in the FAJA belt, Venezuela. Alternatively, oil may be produced with sand, but without solution gas in still other areas.

It is believed that CHOPS production occurs with the formation of wormholes, tunnels that may extend some distance into the formation. There are no current methods for predicting the distribution, location, length, or diameter of wormholes, and there are very limited means of measuring them once formed. Surface seismic surveys may give an indication of their distribution and density. Hence, there is considerable uncertainty about the behavior of CHOPS wells.

CHOPS wells are vertical or slightly deviated wells. They are cased and perforated, and a downhole pump is deployed to create an aggressive pressure differential between formation and wellbore pressures. This causes natural gas to break out of solution from the heavy oil, resulting in foamy oil. Gas bubbles evolving at the wormhole-sand interface destabilize sand grains, and the expanding gas helps move the mixture through the wormholes. Gravity drive on the unconsolidated sands also provides energy for production. At the start-up of production, up to 10% sand by volume is produced along with oil, water, and gas. Sand production eventually falls to under 2% during the well lifetime.

The recovery factor for CHOPS wells is low, typically less than 10%. Therefore, the well must be drilled, completed, and operated as economically as possible.

A large number of CHOPS wells are shut in, having stopped producing oil. There are many possible reasons. A well may water-out if a wormhole reaches a water zone since water flows preferentially due to its much lower viscosity. The wormhole may reach a region where there is insufficient natural gas in the oil to break off sand grains. The sand face may become too strong to allow the wormhole to grow. The wormhole may migrate and interact with wormholes from other wells that have stopped producing. The wormhole may collapse. Infill drilling a well for CHOPS may result in a nonproductive well surrounded by nearby productive wells. Alternatively, an infill well may encounter severe lost-circulation problems if it intercepts an existing wormhole.

The surface footprint for CHOPS wells is small; it only requires space for the wellhead, a storage tank, and a small doghouse. Any produced gas is used on site to power equipment or to heat the storage tank. Because a large volume of sand is produced, pipelines cannot be used for transportation. Instead, trucks are required to move oil, water, and sand for processing or disposal. During spring break-up, the CHOPS wells in Alberta must be shut in since trucks cannot navigate the roads.

The technical challenges for using CHOPS wells include gaining a better understanding of their behavior and developing more predictive performance models.

A major breakthrough would be a secondary recovery method to tap the remaining approximately 90% of the original oil in place. Two possible secondary recovery methods are in situ combustion and solvent flooding. In addition, a primary production method that has a higher recovery factor would also have a significant impact. A high recovery factor and oil production without sand might replace trucking with pipelines, thus reducing carbon dioxide emissions as well as manpower costs and allowing year-around production.

6.2.4 Pressure Pulse Technology

Pressure pulse technology (PPT) is a technology that can be used to enhance the recovery rate of nonaqueous phase liquid (NAPL) and to reduce solids clogging in wells, permeable reactive barriers, and fractured media (CRA, 2003). This technology is based on the discovery that large amplitude pressure pulses that are dominated by low-frequency wave energy generate enhanced flow rates in porous media. For example, in preliminary experiments in heavy oil reservoirs in Alberta, pressure pulse technology has reduced the rate of depletion, increased the oil recovery ratio, and prolonged the life of wells.

The technology uses steady, nonseismic pulse vibrations (e.g., 15 pulses per minute) that generate a low velocity wave effect to encourage flow of oils and small solid particles. It is effective in geologic formations exhibiting elastic properties, such as unconsolidated sediments and sedimentary rocks. It must be applied in a downhole manner in order to be effective. It has been used by the oil industry to improve oil recovery from otherwise exhausted reserves for many years. Also, it has been found that very large amplitude pressure pulses applied for five to 30 hours to a blocked producing well can re-establish economic production in a CHOPS well for many months, even years. Pulsing has been applied in injector wells for improving the efficiency of waterflood patterns and has shown indications of increased oil production and decreased water cut. Additional potential applications include improving the effectiveness of matrix acidizing and diversion (Halliburton, 2004; Mok, 2004).

The mechanism by which pressure pulse technology works is the generation of a porosity dilation wave (a fluid displacement wave similar to a tidal wave). This produces pore-scale dilation and contraction so that oil and water flow into and out of pores, leading to periodic fluid accelerations in the pore throats. As the porosity dilation wave moves through the porous medium at a velocity of about 50 to 100 feet per second (40 to 80 meters per second), the small expansion and contraction of the pores with the passage of each packet of wave energy helps unblock pore throats; increase the velocity of liquid flow; overcome part of the effects of capillary blockage; and reduce some of the negative effects of instability due to viscous fingering, coning, and permeability streak channeling.

Although a very new concept (dating only from 1999 in small-scale field experiments), pressure pulse technology promises to be a major adjunct to a number of oil production processes, particularly all pressure-driven processes, where it will both accelerate flow rates as well as increase oil recovery factors. It is also now used in environmental applications to help purge shallow aquifers of nonmiscible phases such as oil, with about six successful case histories to date.

6.2.5 Solvent Processes

Solvent-based methods have been developed to move heavy oil. A diluent such as naphtha or light oil may be injected near the pump to reduce the viscosity of the heavy oil and allow it to be more easily pumped. Alternatively, diluent may be added at the surface to facilitate pipeline transport.

Vapor-assisted petroleum extraction (VAPEX) is a nonthermal, solvent-based, relatively cold (40°C), low pressure process in which two parallel horizontal wells are drilled with about a 15-feet vertical separation (Yazdani and Maini, 2008). The process involves the injection of vaporized solvents such as ethane, propane, butane, and naphtha to create a vapor chamber (Butler and Mokrys, 1991; Butler and Mokrys, 1995 [United States Patent 5,407,009; United States Patent 5,607,016], Butler and Jiang, 2000). The vapor travels to the oil face where it condenses into a liquid, and the solvent mixed with the oil flows to the lower well (gravity drainage) and is pumped to the surface.

The process can be applied in paired horizontal wells, single horizontal wells, or a combination of vertical and horizontal wells. The physics of the VAPEX process are essentially the same as for the SAGD process (Chapter 7), and the configuration of wells is generally similar. The key benefits are claimed to be (1) significantly lower energy costs, (2) the potential for *in situ* upgrading, and (3) the application to thin reservoirs, with bottom water or reactive mineralogy.

In addition to concern over the cost of the solvent, there are the usual concerns about the interaction of the solvent with the reservoir minerals. Clay is known to adsorb organic solvents very strongly—and the integrity of the reservoir formations and associated strata—a minor fault can cause loss of the solvent as well as cause environmental havoc in underground formations such as aquifers. There is also the concern over the deposition of asphaltic material and its effect on reservoir permeability. Although the function of the solvents might be to extract soluble components of heavy oil, initial contact between the solvent and the oil at a low solvent-to-oil ratio will cause solubilization of the asphaltic constituents (Mitchell and Speight, 1973) and later deposition of these constituents as the solvent-to-oil ratio increases in the later stages of the process.

Because of the slow diffusion of gases and liquids into viscous oils, this approach, used alone, perhaps will be suited only for less viscous oils. Preliminary tests indicate, however, that there are micromechanisms that act so that the VAPEX dilution process is not diffusion-rate limited, and the process may be suitable for the highly viscous heavy oil and tar sand bitumen (Yang and Gu, 2005a, 2005b).

6.3 Enhanced Oil Recovery Methods

Tertiary recovery begins when secondary oil recovery is not enough to continue adequate production, but only when the oil can still be extracted profitably. This depends on the cost of the extraction method and the current price of crude oil. When prices are high, previously unprofitable wells are brought back into production; when they are low, production is curtailed. Tertiary recovery allows another 5% to 15% of the reservoir's oil to be recovered.

Tertiary oil recovery reduces the viscosity of the oil to increase oil production. Thermally enhanced oil recovery (TEOR) methods (Chapter 5) are tertiary recovery techniques that heat the oil and make it easier to extract. Steam injection is the most common form of thermally enhanced oil recovery and is often done with a cogeneration plant. In this type of cogeneration plant, a gas turbine is used to generate electricity, and the waste heat is used to produce steam, which is then injected into the reservoir. In situ combustion is another form of thermally enhanced oil recovery, but instead of steam, some of the oil is burned to heat the surrounding oil. Occasionally, detergents are also used to decrease oil viscosity. Another method is carbon dioxide flooding.

However, there is a renewed interest in chemical enhanced oil recovery (Figure 6–1 and Figure 6–2) because of diminished reserves and advances in surfactant and polymer technology. Greater understanding of the chemical reactions involved has led to good results in the field (Krumrine and Falcone, 1987). Combinations of chemicals may be applied as premixed slugs or in sequence.

The choice of the method and the expected recovery depend on many considerations, economic as well as technological. Only a few recovery methods have been commercially successful, such as steam injection based processes in heavy oils (if the reservoir offers favorable conditions for such applications) and miscible carbon dioxide for light oil reservoirs (Thomas, 2008); other methods remain largely of academic interest. Methods for improving oil recovery, in particular those concerned with lowering the interstitial oil saturation, have received a great deal of attention both in the laboratory and in the field. From the vast amount of literature on the subject, one gets the impression that it is relatively simple to increase oil recovery beyond secondary recovery (assuming that the reservoir lends itself to primary and secondary recovery), but this is not the case (Thomas, 2008).

6.3.1 Alkaline Flooding

Alkaline flooding (*caustic flooding*) (Chapter 5) is an enhanced oil recovery technique in which an alkaline chemical such as sodium hydroxide, sodium orthosilicate, or sodium carbonate is injected during waterflooding or during polymer flooding operations. The alkaline chemical reacts with certain types of oils, forming surfactants inside the reservoir. Eventually, the surfactants reduce the interfacial tension between oil and water and trigger an increase in oil production. However, alkaline flooding is not recommended for carbonate reservoirs because of the abundance of calcium; the mixture between the alkaline chemical and the calcium ions can produce hydroxide precipitation that may damage the formation.

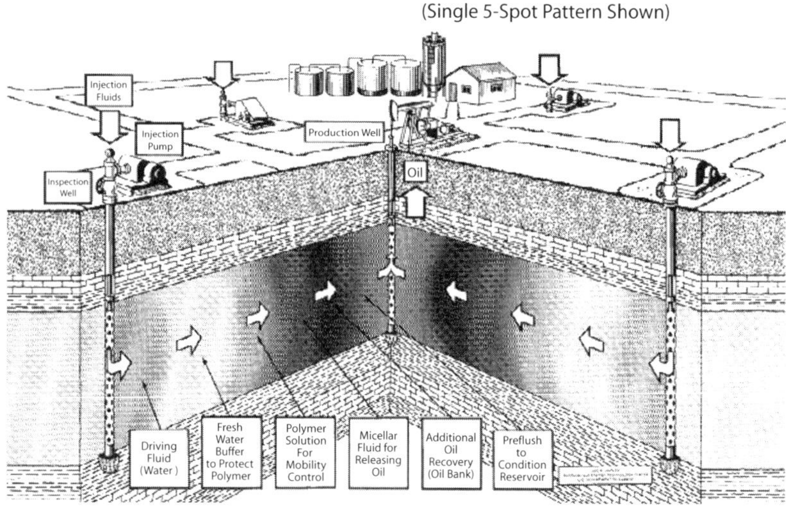

Figure 6–1 *Schematic for Chemical Enhanced Recovery Processes.*

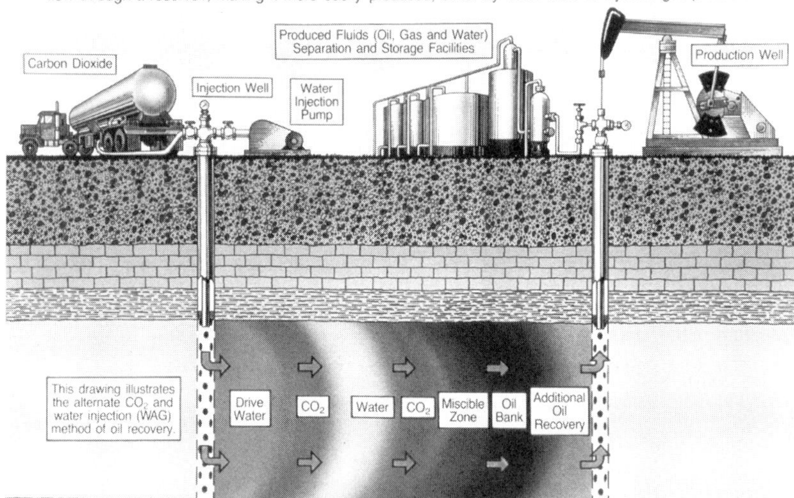

Figure 6–2 *Schematic for Miscible Enhanced Recovery Processes.*

Alkaline flooding is one of several chemical-based enhanced oil recovery (CEOR) methods that are an advancement on conventional secondary waterflooding operations. Chemical-based enhanced oil recovery techniques produce fewer amounts of greenhouse gases than thermal-based enhanced oil recovery approaches. A modification to the process is the addition of surfactant and polymer to the alkali, giving rise to an alkaline-surfactant polymer (ASP) enhanced oil recovery method, essentially a less costly form of micellar-polymer flooding.

The alkaline agents aid in the displacement of crude oil by raising the pH of the injected flood water. The alkali reacts with acidic components in crude oil, forming an in situ surfactant at the oil/brine interface. This mixture then mobilizes the crude oil and removes it from the pore spaces in the reservoir.

Alkaline flooding has been used only in reservoirs containing specific types of high-acid crude oils. Although surfactant flooding is expensive, it has been used in a few large-scale projects and has been demonstrated to have excellent potential for improving the recovery of low-to-moderate-viscosity oils. Chemically-enhanced oil recovery is commercially available under limited conditions, determined by reservoir characteristics, including depth, salinity, and pH. The high cost of chemicals and reservoir characterization studies needs to be reduced to allow expanded use of chemical enhanced oil recovery methods before full commercialization can take place.

The addition of silicates is an enhancement to alkaline flooding. The silicates play two major functions: (1) as a buffer, maintaining a constant high pH level to produce a minimum interfacial tension and (2) improving surfactant efficiency through the removal of hardness ions from reservoir brines, thus reducing adsorption of surfactants on rock surfaces.

6.3.2 Carbon Dioxide Flooding

Carbon dioxide (CO_2) flooding (Chapter 5) is a process whereby carbon dioxide is injected into an oil reservoir in order to increase output when extracting oil. Carbon dioxide flooding is particularly effective in reservoirs deeper than 2,000 ft., where carbon dioxide is in a supercritical state, with the oil gravity on the order of greater than 22° to 25°. Carbon dioxide flooding is not effected by the lithology of the reservoir area but simply by the reservoir characteristics.

First tried in 1972 in Scurry County, Texas, carbon dioxide injection has been used successfully throughout the Permian Basin of West Texas and eastern New Mexico. It is now being pursued to a limited extent in Kansas, Mississippi, Wyoming, Oklahoma, Colorado, Utah, Montana, Alaska, and Pennsylvania.

Carbon dioxide flooding is commonly used to recover oil from reservoirs in which the initial pressure has been depleted through primary production and possibly waterflooding (Orr et al., 1982).

If a well has been produced before and has been designated suitable for carbon dioxide flooding, the first action is to restore the pressure within the reservoir to one suitable for production. This is done by injecting water (with the production well shut off). Once the reservoir is at this pressure, the next step is to inject the carbon dioxide into the same injection wells used to restore pressure. The carbon dioxide gas is forced into the reservoir and is required to come into contact with the oil. This creates a miscible zone that can be moved easier to the production well.

The initial carbon dioxide slug is typically followed by alternate water and carbon dioxide injection. The water serves to improve sweep efficiency and to minimize the amount of carbon dioxide required for the flood. Production is from an oil bank that forms ahead of the miscible front. As reservoir fluids are produced through production wells, the carbon dioxide reverts to a gaseous state and provides a "gas lift" similar to that of original reservoir natural gas pressure.

Until recently, most of the carbon dioxide used for enhanced oil recovery had come from naturally occurring reservoirs. However, new technologies are being developed to produce carbon dioxide from industrial applications such as natural gas processing, fertilizer, ethanol, and hydrogen plants in locations where naturally occurring reservoirs are not available. One demonstration at the Dakota Gasification Company's plant in Beulah, North Dakota, is producing carbon dioxide and delivering it by a new 204-mile pipeline to the Weyburn oil field in Saskatchewan, Canada. EnCana, the field's operator, is injecting the carbon dioxide to extend the field's productive life, hoping to add another 25 years and as many as 130 million barrels of oil that might otherwise have been abandoned.

Additional work has examined possible improvements in carbon dioxide enhanced oil recovery technologies beyond the state of the art that can further increase this potential. This work illustrates that the wide-scale implementation of next-generation technology advances have the potential to increase domestic oil recovery efficiency from about 33% (one third) to over 60%.

The presence of an oil-bearing transition zone beneath the traditionally defined base (oil-water contact) of an oil reservoir is well established. What is now clear is that, under certain geologic and hydrodynamic conditions, an additional residual oil zone (ROZ) exists below this transition zone. This resource could add another 100 billion barrels of oil resource in place in the United States, and an estimated 20 billion barrels could be recoverable with state-of-the-art carbon dioxide enhanced oil recovery technologies.

Large volumes of technically recoverable domestic oil resources remain undeveloped and are yet to be discovered in the United States. Undeveloped domestic oil resources still in the ground (in place) total 1,124 billion barrels. Of this large in-place resource, 430 billion barrels is estimated to be technically recoverable. This resource includes undiscovered oil, "stranded" light oil amenable to carbon dioxide enhanced oil recovery technologies, unconventional oil (deep heavy oil and tar sands), and new petroleum concepts (residual oil in reservoir transition zones).

6.3.3 Cyclic Carbon Dioxide Stimulation

Cyclic carbon dioxide stimulation, also known as the huff-and-puff method, is a single-well operation that is developing as a method of rapidly producing oil. Similar to the cyclic steam process, carbon dioxide is injected into an oil reservoir, the well is shut in for a time (providing for a *soak period*), and then the well is opened, allowing the oil and fluids to be produced. The dissolving of the carbon dioxide in the oil reduces the oil's viscosity and causes it to swell, allowing the oil to flow more easily toward the well. The process can also be used in heavy oil reservoirs by high-pressure injection of carbon dioxide to facilitate miscibility between the oil and carbon dioxide and in cases where thermal methods are not feasible.

Miscible carbon dioxide enhanced oil recovery is a multiple contact process, involving the injected carbon dioxide and the reservoir's oil. During this multiple contact process, carbon dioxide will vaporize the lighter oil fractions into the injected carbon dioxide phase, and carbon dioxide will condense into the reservoir's oil phase. This leads to two reservoir fluids that become miscible (mixing in all parts), with favorable properties of low viscosity, a mobile fluid, and low interfacial tension. The primary objective of miscible carbon dioxide-enhanced oil recovery is to remobilize and dramatically reduce the after-waterflooding residual oil saturation in the reservoir's pore space.

When insufficient reservoir pressure is available or the reservoir's oil composition is less favorable (heavier), the injected carbon dioxide will not become miscible with the reservoir's oil. Then, another oil displacement mechanism, immiscible carbon dioxide flooding, occurs. The main mechanisms involved in immiscible carbon dioxide flooding are: (1) oil phase swelling, as the oil becomes carbon dioxide saturated with carbon dioxide; (2) viscosity reduction of the swollen oil and carbon dioxide mixture; (3) extraction of lighter hydrocarbon into the carbon dioxide phase; and (4) fluid drive plus pressure.

This combination of mechanisms enables a portion of the reservoir's remaining oil to be mobilized and produced. In general, immiscible carbon dioxide enhanced oil recovery is less efficient than miscible carbon dioxide enhanced oil recovery.

6.3.4 Nitrogen Flooding

Nitrogen flooding can be used to recover light oil that is capable of absorbing added gas under reservoir conditions. When nitrogen is injected into a reservoir, it forms a miscible front by vaporizing lighter oil components. As the front moves away from the injection wells, its leading edge goes into solution (or becomes miscible) with the reservoir oil. Continued injection moves the bank of displaced oil toward production wells. Water slugs are injected alternately with the nitrogen to increase the sweep efficiency and oil recovery. Nitrogen can be manufactured on site at relatively low cost by extraction from air by cryogenic separation, and being totally inert it is noncorrosive.

Because of its lower cost, in *nitrogen-carbon dioxide flooding,* the nitrogen can be used in a carbon dioxide flood to displace the carbon dioxide slug and its oil bank.

6.3.5 Polymer Flooding

Polymer flooding is an enhanced oil recovery method (Chapter 5) that uses polymer solutions to increase oil recovery by increasing the viscosity of the displacing water to decrease the water/oil mobility ratio. Polymer flooding is used under certain reservoir conditions that lower the efficiency of a regular waterflood, such as fractures or high-permeability regions that channel or redirect the flow of injected water or heavy oil that is resistant to flow. Adding a water-soluble polymer to the waterflood allows the water to move through more of the reservoir rock, resulting in a larger percentage of oil recovery. Polymer gel is also used to shut off high-permeability zones. In the process, the volumetric sweep is improved, and the oil is more effectively produced. Often, injectivity is one of the critical factors. The polymer solution should therefore be a non-Newtonian and shear thinning fluid, i.e., the viscosity of the fluid decreases with increasing shear rate.

There are three potential ways in which polymer flooding makes the oil recovery process more efficient: (1) through the effects of polymers on fractional flow, (2) by decreasing the water/oil mobility ratio, and (3) by diverting injected water from zones that have been swept. The most important preconditions for polymer flooding are reservoir temperature and the chemical properties of reservoir water. At high temperature or with high salinity in reservoir water, the polymer cannot be kept stabile, and polymer concentration will lose most of its viscosity.

6.3.6 Micellar Polymer Flooding

The *micellar polymer flooding* method uses the injection of a micellar slug into a reservoir. The micellar slug contains a mixture of a surfactant, cosurfactant, alcohol, brine, and oil that moves through the oil-bearing formation, releasing much of the oil trapped in the rock. This method is one of the most efficient enhanced oil recovery methods, but is also one of the most costly to implement.

The slug acts to release oil from the pores of the reservoir rock much like a dishwashing detergent releases grease from dishes so that it can be flushed away by flowing water. To further enhance production, polymer-thickened water for mobility control (as described in the polymer flooding process) is injected behind the micellar slug. Here again, a buffer of fresh water is normally injected following the polymer and ahead of the drive water to prevent contamination of the chemical solutions.

6.3.7 Microbial Enhanced Oil Recovery

Microbial enhanced oil recovery (*MEOR*) processes involve the use of reservoir microorganisms or specially selected natural bacteria to produce specific metabolic events that lead to enhanced oil recovery. The processes that facilitate oil production are complex and may involve multiple biochemical processes. Microbial biomass or biopolymers may plug high-permeability zones and lead to a redirection of the water flood, may produce surfactants that lead to increased mobilization of residual oil, may increase gas pressure by the production of carbon dioxide, or may reduce the oil viscosity due to digestion of large molecules (Banat, 1995; Clark et al., 1981; Stosur, 1991).

From a microbiologist's perspective, microbial enhanced oil recovery processes are somewhat akin to in situ bioremediation processes. Injected nutrients, together with indigenous or added microbes, promote in situ microbial growth and/or generation of products that mobilize additional oil and move it to producing wells through reservoir repressurization, interfacial tension/oil viscosity reduction, and selective plugging of the most permeable zones (Bryant et al., 1989; Bryant and Lindsey, 1996). Alternatively, the oil-mobilizing microbial products may be produced by fermentation and injected into the reservoir.

This technology requires consideration of the physicochemical properties of the reservoir in terms of salinity, pH, temperature, pressure, and nutrient availability (Khire and Khan. 1994a, 1994b). Only bacteria are considered promising candidates for microbial enhanced oil recovery. Molds, yeasts, algae, and protozoa are not suitable due to their size or inability to grow under the conditions present in reservoirs. Many petroleum reservoirs have high concentrations of sodium chloride (Jenneman, 1989) and require the use of bacteria that can tolerate these conditions is necessary (Shennan and Levi, 1987). Bacteria producing biosurfactants and biopolymers can grow at sodium concentrations up to 8% and selectively plug sandstone to create a biowall to recover additional oil (Raiders et al., 1989).

Organisms that participate in oil recovery produce a variety of fermentation products such as carbon dioxide, methane, hydrogen, biosurfactants, and hydrocarbons. Organic acids produced through fermentation readily dissolve carbonates and can greatly enhance permeability in limestone reservoirs. Attempts have been made to promote anaerobic production.

The microbial enhanced oil recovery process may modify the immediate reservoir environment in a number of ways that could also damage the production hardware or the formation itself. Certain sulfate reducers can produce H_2S, which can corrode pipeline and other components of the recovery equipment.

Despite numerous microbial enhanced oil recovery tests, considerable uncertainty remains regarding process performance. Ensuring success requires an ability to manipulate environmental conditions to promote growth and/or product formation by the participating microorganisms. Exerting such control over the microbial system in the subsurface is itself a serious challenge. In addition, conditions vary from reservoir to reservoir, which calls for reservoir-specific customization of the microbial enhanced oil recovery process. This alone has the potential to undermine the economic viability of the microbial process.

Microbial enhanced oil recovery differs from chemical enhanced oil recovery in the method by which the enhancing products are introduced into the reservoir. In oil recovery by the *cyclic microbial method*, a solution of nutrients and microorganisms is introduced into the reservoir during injection. The injection well is shut for an incubation period to allow the microorganisms to produce carbon dioxide gas and surfactants that assist in mobilization of the oil. The well is then opened, and oil and oil products resulting from the treatment are produced. The process is repeated as often as oil can be produced from the well. In oil recovery by *microbial flooding*, the reservoir is usually conditioned by a water flush, and a solution of microorganisms and nutrients is injected into the formation. As this solution is pushed trough the reservoir by water drive, gases and surfactants are formed, and the oil is mobilized and pumped through the well. However, even though microbes produce the necessary chemical reactions in situ whereas surface injected chemicals may tend to follow areas of higher permeability, resulting in decreased sweep efficiency, there is need for caution and astute observation of the effects of the microorganisms on the reservoir chemistry.

In a microbial enhanced oil recovery process, conditions for microbial metabolism are supported via injection of nutrients. In some processes, this involves injecting a fermentable carbohydrate into the reservoir. Some reservoirs also require inorganic nutrients to serve as substrates for cellular growth or to serve as alternative electron acceptors in place of oxygen or carbohydrates.

The stimulation of oil production by in situ bacterial fermentation is thought to proceed by one or a combination of the following mechanisms:

- Improvement of the relative mobility of oil to water by biosurfactants and biopolymers.
- Partial repressurization of the reservoir by methane and carbon dioxide.
- Reduction of oil viscosity through the dissolution of organic solvents in the oil phase.
- Increase of reservoir permeability and widening of the fissures and channels through the etching of carbonaceous rocks in limestone reservoirs by organic acids produced by anaerobic bacteria.
- Cleaning of the well bore region through the acids and gas from in situ fermentation. The gas pushes oil from dead space and dislodges debris that plugs the pores. The average pore size is increased, and as a result, the capillary pressure near the well bore is made more favorable for the flow of oil.
- Selective plugging of highly permeable zones by injecting slime-forming bacteria followed by sucrose solution, which initiates the production of extracellular slimes and improves aerial sweep efficiency.

One of the major attributes of microbial enhanced oil recovery is its low cost, but there must be recognition that it is a single process. Furthermore, reports on the deleterious activities of microorganisms in the oil field contribute to the skepticism of employing technologies using microorganisms. It is also clear that scientific knowledge of the fundamentals of microbiology must be coupled with an understanding of the geological and engineering aspects of oil production in order to develop microbial enhanced oil recovery technology.

Finally, recent developments in *upgrading* of heavy oil and bitumen (Speight, 2007) indicate that the near future could see a reduction of the differential cost of upgrading heavy oil. These processes are based on a better understanding of the issues of asphaltene solubility effects at high temperatures, incorporation of a catalyst that is chemically precipitated internally during the upgrading, and improvement of hydrogen addition or carbon rejection.

6.4 Oil Mining

Oil mining is the term applied to the surface or subsurface excavation of petroleum-bearing formations for subsequent removal of the heavy oil or bitumen by washing, flotation, or retorting treatments. Oil mining also includes recovery of heavy oil by drainage from reservoir beds to mine shafts or other openings driven into the rock or by drainage from the reservoir rock into mine openings driven outside the tar sand but connected with it by bore holes or mine wells.

Oil mining is not new. Mining of petroleum and bitumen has occurred in the Sinai Peninsula, in the Euphrates valley, and in Persia prior to 5000 BC. In addition, subsurface oil mining was used in the Pechelbronn oil field in Alsace, France as early as 1735. This early mining involved the sinking of shafts to the reservoir rock, only 100 to 200 feet (30 to 60 meters) below the surface, and the excavation of the tar sand (oil sand) in short drifts driven from the shafts. These tar sands were hoisted to the surface and washed with boiling water to release the bitumen. The drifts were extended as far as natural ventilation permitted. When these limits were reached, the pillars were removed, and the openings filled with waste. This type of mining continued at Pechelbronn until 1866, when it was found that oil could be recovered from deeper, and more prolific, sands by letting it drain in place through mine openings with no removal of sand to the surface for treatment. Nevertheless, mining for petroleum continues to be a challenge

Oil mining methods should be applied in reservoirs that have significant residual oil saturation and reservoir or fluid properties that make production by conventional methods inefficient or impossible. The high well density in improved oil mining usually compensates for the inefficient production caused by reservoir heterogeneity. However, close well spacing can also magnify the deleterious effects of reservoir heterogeneity. If a high-permeability streak exists with a lateral extent that is less than the interwell spacing of conventional wells but is comparable to that of improved oil mining, the channeling is more unfavorable for the improved oil mining method.

Engineering a successful oil mining project must address a number of items: (1) there must be sufficient recoverable resources, (2) the project must be conducted safely, and (3) the project should be engineered to maximize recovery within economic limits. The use of a reliable screening technique is necessary to locate viable candidates. Once the candidate is defined, an exhaustive literature search should follow, covering the local geology, drilling, production, completion, and secondary and tertiary recovery operations.

The reservoir properties, which can affect the efficiency of heavy oil or bitumen production by mining technology, can be grouped into three classes:

1. *Primary properties* are those properties that influence fluid flow and fluid storage properties. They include rock and fluid properties, such as porosity, permeability, wettability, crude oil viscosity, and pour point.

2. *Secondary properties* are those properties that significantly influence the primary properties, including pore size distribution, clay type, and content.

3. *Tertiary properties* are those other properties that mainly influence oil production operations (fracture breakdown pressure, hardness, and thermal properties) and mining operations (e.g., temperature, subsidence potential, and fault distribution).

There are also important rock mechanical parameters of the formation in which a tunnel is to be mined and from where all oil mining operations will be conducted. These properties are mostly related to the mining aspects of the operations, and not all are of equal importance in their influence on the mining technology. Their relative importance also depends on the individual reservoir.

Many of the candidate reservoirs for application of improved oil mining are those with high oil saturation resulting from the adverse effects of reservoir heterogeneity. Faulting, fracturing, and barriers to fluid flow are features that cause production of shallow reservoirs by conventional methods to be inefficient. Production of heterogeneous reservoirs by underground oil production methods requires consideration of the manner in which fractures alter the flow of fluids.

In a highly fractured formation with low matrix permeability, the fluid conductivity of the fracture system may be much more than that of the matrix rock. In a highly fractured reservoir with low matrix permeability and reasonably high porosity, the fracture system provides the highest permeability to the flow of oil, but the matrix rock contains the greater volume of the oil in place. The rate of the flow of oil from the matrix rock into the fracture system, the extent and continuity of the fracture system, and the degree to which the production wells effectively intersect the fracture system determine the production rate. Special consideration must be given to these factors in predicting production rates in fractured reservoirs. Under favorable circumstances, higher production rates may be achieved in fractured reservoirs by improving mining methods. Other reservoirs that are good candidates for oil mining are those that are shallow, have high oil saturation, have a nearby formation that is competent enough to support the mine, and cannot be efficiently produced by conventional methods.

Surface mining is the mining method that is currently being used by Suncor Energy and Syncrude Canada Limited to recover tar sand from the ground. Surface mining can be used in mineable tar sand areas that lie under 75 meters (250 feet) or fewer of overburden material. Only 7% of the Athabasca Oil Sands deposit can be mined using the surface mining technique; the other 93% of the deposit has more than 75 meters of overburden. This other 93% has to be mined using different mining techniques.

The first step in surface mining is the removal of muskeg and overburden. Muskeg is a water-soaked area of decaying plant material that is one to three meters thick and lies on top of the overburden material. Before the muskeg can be removed, it must be drained of its water content. The process can take up to three years to complete. Once the muskeg has been drained and removed, the overburden must also be removed. Overburden is a layer of clay, sand, and silt that lies directly above the tar sand deposit. Overburden is used to build dams and dykes around the mine and is eventually be used for land reclamation projects. When all of the overburden is removed, the tar sand is exposed and ready to be mined.

There are two methods of mining currently in use in the Athabasca Oil Sands. Suncor Energy uses the truck-and-shovel method of mining, whereas Syncrude uses the truck-and-shovel method along with draglines and bucket-wheel reclaimers. These enormous draglines and bucket-wheels are being phased out and soon will be completely replaced with large trucks and shovels. The shovel scoops up the tar sand and dumps it into a heavy hauler truck. The heavy hauler truck takes the tar sand to a conveyor belt that transports the tar sand from the mine to the extraction plant. Presently, there are extensive conveyor belt systems that transport the mined tar sand. With the development of new technologies, these conveyors are being phased out and replaced with hydrotransport technology. Hydrotransport is a combination of ore transport and preliminary extraction. After the bituminous sand has been recovered using the truck and shovel method, it is mixed with water and caustic soda to form a slurry. This is pumped along a pipeline to the extraction plant. The extraction process thus begins with the mixing with water and agitation needed to initiate bitumen separation from the sand and clay.

Mine spoils need to be disposed of in a manner that assures physical stabilization, meaning appropriate slope stability for the pile against not only gravity but also earthquake forces. Since return of the spoils to the mine excavations is seldom economical, the spoil pile must be designed as a permanent structure whose outline blends into the landscape. Straight, even lines in the pile must be avoided.

Even though estimates of the recoverable oil from the Athabasca deposits are only on the order of 27 by 109 bbl of synthetic crude oil (representing <10% of the total in-place material), this is, for the Canadian scenario, approximately six times the estimated volume of recoverable conventional crude oil. In addition, the comparative infancy of the development of alternative options almost ensured the adoption of the mining option for the first two (and even later) commercial ventures in Athabasca.

Underground mining options have also been proposed but, for the moment, have not been developed because of the fear of collapse of the formation onto any operation/equipment. This particular option should not, however, be rejected out of hand. A novel aspect or the requirements of the developer (which remove the accompanying dangers) may make such an option acceptable. Currently, bitumen is recovered commercially from tar sand deposits by a mining technique. This produces tar sand that is sent to the processing plant for separation of the bitumen from the sand prior to upgrading.

The oil mining method of recovery has received considerable attention since it was chosen as the technique of preference for the only two commercial bitumen recovery plants in operation in North America. In situ processes have been tested many times in the United States, Canada, and other parts of the world and are ready for commercialization. There are also conceptual schemes that are a combination of both mining (above-ground recovery) and in situ (nonmining recovery) methods.

The bitumen occurring in tar sand deposits poses a major recovery problem. The material is notoriously immobile at formation temperatures and must therefore require some stimulation (usually by thermal means) in order to ensure recovery. Alternately, proposals have been noted that advocate bitumen recovery by solvent flooding or by the use of emulsifiers. There is no doubt that, with time, one or more of these functions may come to fruition; for the present, the two commercial operations rely on the mining technique.

There are two methods of mining currently in use in the Athabasca Tar Sands deposits. Suncor Energy uses the truck and shovel method of mining whereas Syncrude uses the truck and shovel method of mining, as well as draglines and bucket-wheel reclaimers. These enormous draglines and bucket-wheels are being phased out and soon will be completely replaced with large trucks and shovels. The shovel scoops up the tar sand and dumps it into a heavy hauler truck. The heavy hauler truck takes the tar sand to a conveyor belt that transports the tar sand from the mine to the extraction plant. Presently, there are extensive conveyor belt systems that transport the mined tar sand from the recovery site to the extraction plant. With the development of new technologies these conveyors are being phased out and replaced with hydrotransport technology. Hydrotransport is a combination of ore transport and preliminary extraction. After the bituminous sands have been recovered using the truck and shovel method, it is mixed with water and caustic soda to form a slurry and is pumped along a pipeline to the extraction plant. The extraction process thus begins with the mixing of the water and agitation needed to initiate bitumen separation from the sand and clay.

The *hot-water process* (Clark, 1944) is, to date, the only successful commercial process to be applied to bitumen recovery from mined tar sands in North America (Clark, 1944; Carrigy, 1963a; Carrigy 1963b; Fear and Innes, 1967; Speight and Moschopedis, 1978). Many process options have been tested with varying degrees of success, and one of these options may even supersede the hot-water process.

The process utilizes the linear variation of bitumen density and the nonlinear variation of water density with temperature so that the bitumen, which is heavier than water at room temperature, becomes lighter than water at 80°C (180°F). Surface-active materials in the tar sand also contribute to the process. The essentials of the hot-water process involve conditioning, separation, and scavenging.

In the hot-water extraction process, the tar sand feed is introduced into a *conditioning* drum. In this step, the tar sand is heated and mixed with water, and agglomeration of the oil particles begins. The conditioning is carried out in a slowly rotating drum that contains a steam-sparging system for temperature control as well as mixing devices to assist in lump-size reduction and a size ejector at the outlet end. The tar sand lumps are reduced in size by ablation and mixing action.

In the conditioning step, also referred to as mixing or pulping, tar sand feed is heated and mixed with water to form a pulp of 60% by weight to 85% by weight solids at 80°C to 90°C (175°F to 196°F). First the lumps of tar sand as-mined are reduced in size by ablation, i.e., successive layers of lump are warmed and sloughed off revealing cooler layers. The conditioned *pulp*, which has the following characteristics: (1) 60% to 85% by weight solids and (2) a 7.5 to 8.5 pH, is screened through a double-layer vibrating screen. Water is then added to the screened material (to achieve more beneficial pumping conditions), and the pulp enters the separation cell through a central feed well and distributor. The bulk of the sand settles in the cell and is removed from the bottom as tailing, while the majority of the bitumen floats to the surface and is removed as froth. A middlings stream (mostly of water with suspended fines and some bitumen) is withdrawn from approximately midway up the side of the cell wall. Some of the middlings are recycled to dilute the conditioning drum effluent for pumping. Clays do not settle readily and generally accumulate in the middlings layer.

The combined froth from the separation cell and scavenging operation contains an average of about 10% by weight mineral material and up to 40% by weight water. Demineralizing and dewaterizing is accomplished in two stages of centrifuging. In the first stage, the coarser mineral material is removed but much of the water remains. The feed then passes through a filter to remove any additional large-size mineral matter that would plug up the nozzles of the second-stage centrifuges.

Bituminous froth from the hot-water process may be mixed with a hydrocarbon diluent (e.g., coker naphtha) and centrifuged. The Suncor process employs a two-stage centrifuging operation; each stage consists of multiple centrifuges of conventional design installed in parallel. The bitumen product contains 1% to 2% by weight mineral (dry bitumen basis) and 5% to 15% by weight water (wet diluted basis). Syncrude also utilizes a centrifuge system with naphtha diluent.

The bitumen isolated by the hot-water separation process is then sent for upgrading to a synthetic crude oil.

Environmental regulations in Canada and the United States will not allow the discharge of tailing streams into a river, onto the surface, or onto any area where groundwater domains or a river may be contaminated. The tailing stream is essentially high in clays and contains some bitumen, hence the current need for tailings ponds, where some settling of the clay occurs. In addition, an approach to acceptable reclamation of the tailings ponds will have to be accommodated at the time of site abandonment.

6.5 References

Banat, I. M. 1995. Biosurfactant Production and Possible Uses in Microbial Enhanced Oil Recovery and Oil Pollution Remediation. Bioresearch Technology. 51:1–12.

Butler, R.M., and Mokrys, I.J. 1991. Journal of Canadian Petroleum Technology. 30(1):97–106.

Butler, R.M., and Mokrys, I.J. 1995. Process and Apparatus for the Recovery of Hydrocarbons from a Hydrocarbon Deposit. United States Patent 5,407,009.

Butler, R.M., and Mokrys, I.J. 1995. Process and Apparatus for the Recovery of Hydrocarbons from a Hydrocarbon Deposit. United States Patent 5,607,016.

Butler, R.M., and Jiang, Q. 2000. Journal of Canadian Petroleum Technology. 39:48–56.

Bryant, R.S., and Lindsey, R.P. 1996. World-wide Applications of Microbial Technology for Improving Oil Recovery, 27–134. In Proceedings of the SPE Symposium on Improved Oil Recovery. Society of Petroleum Engineers, Richardson, Texas.

Bryant, R. S., Donaldson, E.C., Yen, T.F., and Chilingarian, G.V. 1989. Microbial Enhanced Oil Recovery,. In Enhanced Oil Recovery II

Processes and Operations, ed. E. C. Donaldson, G. V. Chilingarian, and T. F. Yen, 423–450. Elsevier, Amsterdam, The Netherlands.

Carrigy, M.A. 1963a. Bulletin No. 14. Alberta Research Council, Edmonton, Alberta, Canada.

Carrigy, M.A. 1963b. The Oil Sands of Alberta. Information Series No. 45. Alberta Research Council, Edmonton, Alberta, Canada.

Chen, X. 2006. Heavy Oils, Part 1. SIAM News, Society of Industrial and Applied Mathematics. 39(3).

Chugh, S., Baker, R., Telesford, A., and Zhang, E. 2000. Mainstream Options For Heavy Oil: Part I—Cold Production. Journal of Canadian Petroleum Technology. 39(4):31–39.

Clark, J.B., Munnecke, D.M., and Jenneman, G.E. 1981. In Situ Microbial Enhancement of Oil Production. Developments in Industrial Microbiology. 15:695–701.

Clark, K.A. 1944. Transactions of the Canadian Institute of Mining and Metallurgy. 47:257.

Conestoga-Rovers and Associates. 2003. Pressure Pulse Technology. Innovative Technology Group, Conestoga-Rovers and Associates, Niagara Falls, New York. 3(1). http://www.craworld.com/en/newsevents/resources/inn_2003_jan.pdf.

Curtis, C., Kopper, R., Decoster, E., Guzmán-Garcia, A., Huggins, C., Knauer, L., Minner, M., Kupsch, N., Linares, L.M., Rough, H., and Waite, M. 2002. Heavy-Oil Reservoirs in Oilfield Review. 14(3):42–46.

Etebar S: 1995. Captain Innovative Development Approach," SPE 30369 Society of Petroleum Engineers, Proceedings, Offshore Europe. Conference, Aberdeen, September 5–8.

Fear, J.V.D., and Innes, E.D. 1967. Proceedings. 7th World Petroleum Congress. 3:549.

Halliburton. 2004. http://www.onthewavefront.com/Files/Press/2004/Halliburton_19_ July _04.htm.

Jenneman, G.E. 1989. The Potential for In-situ Microbial Applications. Developments in Petroleum Science. 22:37–74.

Khire, J.M., and Khan, M.I. 1994a. Microbially Enhanced Oil Recovery (MEOR). Part 1. Importance and Mechanisms of Microbial Enhanced Oil Recovery. Enzyme Microbial Technology. 16:170–172.

Khire, J.M., and Khan, M.I. 1994b. Microbially Enhanced Oil Recovery (MEOR). Part 2. Microbes and the Subsurface Environment for Microbial Enhanced Oil Recovery. Enzyme Microbial Technology. 16:258–259.

Kovscek, A.R. 2002. Heavy and Thermal Oil Recovery Production Mechanisms. Quarterly Technical Progress Report, Reporting Period: April 1 through June 30, 2002. DOE Contract Number: DE-FC26-00BC15311. United States Department of Energy, Washington, DC

Krumrine, P.H., and Falcone, J.S. Jr. 1987. Beyond Alkaline Flooding: Design of Complete Chemical Systems. SPE 16280. Society of Petroleum Engineers, Houston, Texas.

Kumar, M. 2006. Heavy Oil Recovery—Recent Developments and Challenges. Los Angeles Monthly Petroleum Technology Forum. L.A. Basin Section, Society of Petroleum Engineers. http://www.laspe.org/petrotech/petrooct10906.html.

Loughead, D.J. 1992. Lloydminster Heavy Oil Production—Why So Unusual?; 9th Annual Heavy Oil and Oil Sands Technology Symposium. Calgary, Alberta, Canada, March 11.

McCaffrey, W.J., and Bowman, R.D. 1991. Recent Successes in Primary Bitumen Production. HOOS Technical Symposium Challenges and Innovation. Annual Technical Meeting, Petroleum Society of the Canadian Institute of Mining.

Meyer, R.F., and Attanasi, E.D. 2003. Fact Sheet 70-03, Heavy Oil and Natural Bitumen—Strategic Petroleum Resources. United States Geological Survey. http://pubs.usgs.gov/fs/fs070-03/.

Mitchell, D.L., and Speight, J.G. 1973. The Solubility of Asphaltenes in Hydrocarbon Solvents. Fuel. 52:149.

Mok, J. 2004. http://jonmok.com/OilSands/index.php?option=com_content&task= view&id=18&Itemid=44.

Orr, F.M. Jr., Heller, J.P., and Taber, J.J. 1982. Carbon Dioxide Flooding for Enhanced Oil Recovery, Promise and Problems. Journal of the American Oil Chemists Society. 59(10).

Raiders, R.A., Knapp, R.M., and McInerney, M.J. 1989. Microbial Selective Plugging and Enhanced Oil Recovery. Journal of Industrial Microbiology. 4:215–230.

Safinya, K. 2008. Heavy Oil Recovery—The Road Ahead. Alberta Oil. 4(1):14–19.

Shennan, J.L., and Levi, J.D. 1987. In Situ Microbial Enhanced Oil Recovery. In Biosurfactants and Biotechnology, ed. N. Kosaric, W. L. Cairns, and N. C. C. Gray, p. 163–180. Marcel Dekker, New York.

Speight. J.G., and Moschopedis, S.E. 1978. Factors Affecting Bitumen Recovery by the hot-water Process. Fuel Processing Technology. 1:261.

Speight. J.G. 2007. The Chemistry and Technology of Petroleum. 4th Edition. CRC Press, Taylor and Francis Group, Boca Raton, Florida.

Stosur, G.J. 1991. Unconventional EOR Concepts. Critical Reports In Applied Chemistry. 33:341–373.

Thomas, S. 2008. Enhanced Oil Recovery: An Overview. Oil & Gas Science and Technology—Revue Institut Français du Pétrole. 63(1):9–19.

Yang, C., Gu, Y. 2005a A Novel Experimental Technique for Studying Solvent Mass Transfer and Oil Swelling Effect in a Vapor Extraction (VAPEX) Process. Paper No. 2005-099. Proceedings. 56th Annual Technical Meeting. The Canadian International Petroleum Conference. Calgary, Alberta, Canada, June 7–9.

Yang, C., and Gu, Y. 2005b. Effects of Solvent-Heavy Oil Interfacial Tension on Gravity Drainage in the VAPEX Process. Paper No. SPE 97906. Society of Petroleum Engineers International Thermal Operations and Heavy Oil Symposium. Calgary, Alberta, Canada, November 1–3.

Yazdani, A., and Maini, B.B. 2008. Modeling the VAPEX Process in a Very Large Physical Model. Energy & Fuels. 22:535–544.

CHAPTER 7

THERMAL METHODS OF RECOVERY

Heavy oil is characterized by high viscosity (i.e., resistance to flow measured in centipoises or cp) and high densities compared to conventional oil (IEA, 2005). Although there are many definitions for heavy oil, the most appropriate definition is drawn from United States Federal Energy Administration (FE-76-4), which relates to tar sand

> *...the several rock types that contain an extremely viscous hydrocarbon which is not recoverable in its natural state by conventional oil well production methods including currently used enhanced recovery techniques. The hydrocarbon-bearing rocks are variously known as bitumen-rocks oil, impregnated rocks, oil sands, and rock asphalt.*

By inference, petroleum and heavy oil are recoverable by well production methods and currently used enhanced recovery techniques. For convenience, it is assumed that before depletion of the reservoir energy, conventional crude oil is produced by primary and secondary techniques, whereas heavy oil requires tertiary (enhanced) oil recovery techniques. While this is an oversimplification, it may be used as a general guide. The term *natural state* cannot be defined out of context; in the context of Federal Energy Administration (FEA) Ruling 1976-4, the term is defined in regards to the composition of the heavy oil or bitumen. The final determinant of whether or not a reservoir is a tar sand deposit is the character of the viscous phase (bitumen) and the method that is required for recovery.

Using this definition removes reliance upon a single property such as viscosity or API gravity (subject to experimental difference in the analytical method), which varies from country to country. While a single property might serve to generally define heavy oil, it is by no means accurate and should not be taken as a conclusive definition.

In some cases, the majority of subsurface heavy oil will not flow towards a well bore in sufficient quantity to be economically viable. Thermal recovery involves heating up the reservoir, thereby lowering the heavy oil's viscosity and enabling the oil to flow to the well bore. The API gravity of heavy oil is usually less than 20°, depending upon the reservoir, and viscosity is very high at reservoir temperature. Heavy oil viscosity decreases rapidly with increasing temperatures; therefore, external heat may be required for production. High-temperature steam is commonly used to deliver heat to the formation. The steam/oil ratio (SOR) or fuel/oil ratio is an important measure of the energy required to produce heavy oil; a steam/oil ratio of 3 means that three barrels of water (converted to steam) are needed to produce one barrel of oil.

It is an oversimplification to state that thermal recovery processes are successful because they tend to reduce the viscosity of the crude oil; there are several important factors related to oil chemistry and oil rock interactions that play a role in heavy oil recovery (Prats, 1986; Lake and Walsh, 2004). For example, the role of steam distillation and other physical effects in steam-based processes have long been ignored or not understood. It is only when there is a thorough knowledge of such phenomena that heavy oil recovery will be fully understood.

Generally, the thermal recovery processes used today fall into two classes: (1) processes in which a hot fluid such as steam is injected into the reservoir and (2) processes in which heat is generated within the reservoir itself, such as the combustion processes (Figure 7–1). The former are usually the steam-based processes, whereas the latter are processes such as in situ combustion or fireflooding. Thermal recovery processes also can be classified as thermal drives or stimulation treatments. A thermal drive not only reduces the viscosity but also provides a force to increase the flow rates of the oil to the production well. In thermal stimulation treatments, only the reservoir near production wells is heated.

In thermal drives, fluid is injected continuously into a number of injection wells to displace oil and obtain production from other wells. The pressure required to maintain the fluid injection also increases

Figure 7–1 *Oil recovery by thermal methods.*

the driving forces in the reservoir, increasing the flow of crude oil. Driving forces present in the reservoir, such as gravity, solution gas, and natural water drive, effect the improved recovery rates once the flow resistance is reduced. Stimulation treatments also can be combined with thermal drives, in which case the driving forces are both natural and imposed.

However, the successful recovery technique that is applied to one heavy oil reservoir is not necessarily the technique that will guarantee success for another reservoir. General applicability of the techniques is not guaranteed. Caution is advised when applying the knowledge gained from one resource to the issues of another resource. Although the principles may at first sight appear to be the same, the technology must be adaptable.

In fact, each production method must be tailored for the particular resource and its fluid properties. A method that works in one situation may fail utterly in a different one. It is essential that the properties of the resource (in this case, of the reservoir and the oil) be fully understood before selecting a production scheme. Essential properties include the geological setting; the depth, areal extent, and thickness of the resource; oil composition, density, viscosity, and gas content; the presence of bottom water or top gas zones; petrophysical and geomechanical properties such as porosity, permeability, and rock strength; the presence of shale layers; vertical and horizontal permeability; and the variation of these properties across the reservoir.

Thermal processes have been the only practical means of improving the oil recovery performance of heavy crude oil reservoirs, and the use of heat in well bores to increase production rates for heavy crude oils has been accepted as viable for several decades. For example, cyclic steam injection has been commercially successful since it was discovered in the early 1960s. Processes using hot-water, hot gas, steam, and various kinds of combustion drives have gained varying degrees of acceptance. Even though the application of thermal recovery processes was initiated because of the difficulties encountered in recovering heavy crude oil (specifically, the difficulties in encouraging flow to a production well), the drive processes can also be applied to recovery of the residual oil in energy-depleted reservoirs that hold considerable amounts of conventional (low-viscosity) crude oil.

Thermal decomposition (visbreaking or thermal cracking) can also be used to decrease the viscosity of heavy crude oil. The intramolecular balance of the crude oil is changed with the resulting deposition of a carbonaceous residue that separates from the oil, leaving a lighter (easier-flowing) oil product phase. Visbreaking is a surface process (Speight, 2007), and its chemistry offers opportunities for thermal recovery of heavy oil. The partial upgrade of oil in situ with the deposition of the carbonaceous residue, which contains most of the nitrogen and metals originally in the oil, and the production of partially upgraded oil at the surface is to be preferred. However, the disadvantage is that the deposition of a carbonaceous solid has the potential to result in blockage of the pores and flow channels within the reservoir. Such interference with the recovery process is to be avoided.

It is generally the case that crude oil alteration occurs in all thermal recovery processes, but a significant amount of conversion by thermal changes occurs only at the high temperatures encountered during combustion processes.

7.1 Hot-Fluid Injection

Hot-fluid injection processes are those processes in which preheated fluids are injected into a relatively cold reservoir (Figure 7–1). The injected fluids are usually heated at the surface, although well bore heaters (downhole heaters) are seeing a wider use. Fluids range from water (both liquid and vapor) and air to others such as natural gas, carbon dioxide, exhaust gases, and even solvents. It should be noted here that if water vapor is injected into the reservoir, the process might be more correctly designated as a steam-based process.

In its simplest form, a hot-water flood (hot-water drive) involves the flow of only two phases: water and oil (Dietz, 1967). Hot-water-flooding is basically a displacement process in which oil is displaced immiscibly by both hot and cold water. Except for temperature effects and the fact that they generally are applied to relatively viscous crude oils, hot-water floods have many elements in common with conventional waterfloods.

Because of the pervasive presence of water in all petroleum reservoirs, displacement by hot-water must occur to some extent in all thermal recovery processes. It is known to contribute to the displacement of oil in the zones downstream of both steam drives and combustion drives. Thus, many elements of the discussion on hot-water drives presented in this chapter are applicable to appropriate regions in other processes.

In the hot-water flood, the leading edge of the injected hot-water loses heat so rapidly that it quickly reaches the initial reservoir temperature, and the oil mobility is that of the unheated oil at the leading edge of the displacement front. On the other hand, the viscosity of the injected hot-water is lower than in conventional waterfloods. Thus, the mobility ratio of the oil ahead of the displacement front and the injected water near the injection well is less favorable in hot-water floods than in conventional waterfloods (which should result in somewhat earlier water breakthrough than hot-water floods). On the other hand, the mobility ratio of the fluids in the heated zones is more favorable in the hot-water floods than in conventional waterfloods. This results in better displacement efficiency from the heated zone and would improve the ultimate recovery even where residual oil saturation does not decrease with increasing temperature.

Typically, in hot-waterflooding, the water is filtered, treated to control corrosion and scale, heated, and if necessary, treated to minimize the swelling of clays in the reservoir. The complex mineralogy of clay minerals (Grim, 1968) must be taken into account wherever clay occurs in a reservoir.

Thermal expansion of oil also contributes to the improved displacement efficiency of thermal projects. Gravity separation usually is accentuated by the increased density difference between water and oil as temperatures increase. In addition, most of the reservoir heat is in the zones from which most of the oil already has been displaced. Indeed, all thermal drives are characterized by the presence of large amounts of heat in oil-depleted portions of the reservoir; this has prompted a number of modifications aimed at scavenging or recycling the heat to improve the efficiency of the process (Lake and Walsh, 2004, Chapter 2).

The amount of oil displaced in a hot-water drive is invariably larger than the amount produced. As discussed in Chapter 4, the oil that is displaced but not produced is stored in unswept portions of the reservoir. In the case of viscous crude oils especially, the mobility ratio between the advancing oil and any gas or water in the reservoir is very favorable. This means that crude oil tends to fill regions of the reservoir initially filled with mobile gas and water before it is produced. Where an oil bank forms, consideration of these effects allows an estimation of the recovery history from estimates of the oil displacement history.

Although there are no simple methods for predicting oil recovery from hot-water floods, an approach is suggested that: (1) is based on conventional waterflood technology, (2) appears to have some of the elements necessary to describe hot-water floods, and (3) considers only the effects of permeability variations and mobility ratio. It is by no means a proven method and is offered merely as an example of how it may be possible to adapt the existing technology from a related field when there is no other method for making the desired estimates of oil recovery.

In addition to cost, the choice of the fluid is controlled by the expected effect of the fluid on the production response of the crude oil as well as on and availability of fluids. For example, seawater may be used for injection into an undersea reservoir where the cost of water delivery to the platform would be prohibitive. On the other hand, choice of injection fluid is governed by the nature of the reservoir, with each reservoir being considered a specific entity subject to the character of the oil and the mineralogy of the reservoir rock (Figure 7–2). For example, steam or hot-water should not be injected without first considering the possible consequences of its effects on reservoirs containing significant amounts of swelling clays.

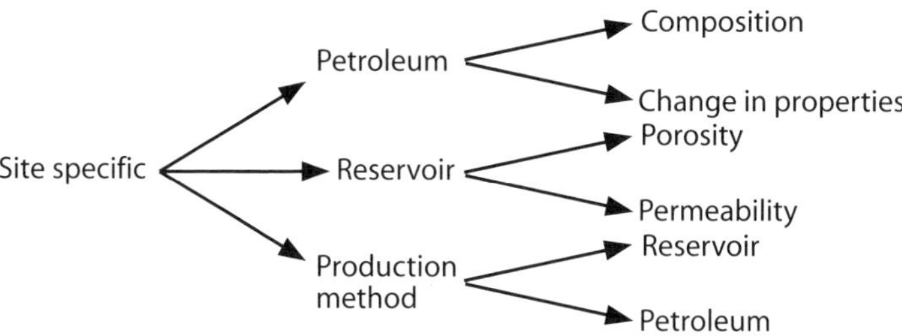

Figure 7–2 *Recovery is site specific and depends upon several variable factors.*

Finally, thermal recovery processes may be used for either stimulation or flooding, and the choice is usually dictated by the properties of the oil and the reservoir. In reservoirs that are small or that have relatively poor continuity, it may not be feasible to drill enough wells to ensure adequate communication to implement a flooding operation. It may be necessary to consider a stimulation treatment to increase both recovery rates and ultimate economic recovery. Cyclic steam injection has proved to be a successful stimulation technique for heavy oil as long as there is the means to produce the crude oil once its viscosity is reduced. Combustion stimulation has been found to be successful in burning solid organic particles, in stabilizing clays, and in increasing absolute permeability near the treated well. These effects are particularly attractive in low-permeability reservoirs.

7.2 Steam-Based Methods

The steam-based processes, as mentioned in Chapter 5, are the most advanced of all enhanced oil recovery methods in terms of field experience and thus have the least uncertainty in estimating performance (EPRI, 1999; Curtis et al., 2002; Advanced Resources International, 2005). However, as with all enhanced oil recovery processes, a good reservoir description is a necessity (Lake et al., 1992). Steam processes are most often applied in reservoirs containing viscous oils and tars, usually in place of rather than following secondary or primary methods. Commercial application of steam processes has been underway since the early 1960s. In contrast, although in situ combustion has been field tested under a wide variety of reservoir conditions, few projects have proven economical and advanced to commercial scale.

In *steam stimulation* processes, heat and drive energy are supplied in the form of steam injected through wells into the heavy oil reservoir. In most instances, the injection pressure must exceed the formation fracture pressure in order to force the steam into the reservoir and into contact with the oil. When sufficient heating has been achieved, the injection wells are closed for a soak period of variable length and then allowed to produce, first applying the pressure created by the injection and then using pumps as the wells cool and production declines.

The concept behind the steam-based processes is generally viscosity reduction so that the heavy oil can flow to the production well. Other effects such as emulsion drive, thermal expansion, solution gas drive, and steam distillation are also believed to occur. The extent of each is unknown and oil dependent. In the steam-stripping processes, steam removes relatively low-boiling components from the heavy oil. In fact, steam stripping removes a larger fraction of the crude oil than would be suggested by the boiling point distribution because of *steam distillation*. There is no chemical change to the constituents of the oil, although there may be favorable compositional changes to the oil insofar as lighter fractions are recovered and heavier materials remain in the reservoir. As the steam condenses, most of the stripped or steam-distilled components in the steam also condense and form a solvent front that will assist in displacing heavy oil (as a solvent/oil mix) towards the production well. These effects help to improve the displacement efficiency.

Whether or not steam distillation occurs and the role it pays in oil recovery depends on the character of the heavy oil as well as the downhole conditions. Steam distillation can indeed augment the process when the steam-distilled material moves with the steam front and acts as a solvent for oil ahead of the steam front (Pratts, 1986).

When dealing with heavy oil, it can be reasonably assumed that the steam-distilled material is composed of aromatics and naphthenic constituents which are excellent solvents for oil (Mitchell and Speight, 1973). On the other hand, if the steam-distilled constituents have a predominantly paraffinic character (which is not considered likely for heavy oil but very likely for residual oil in an energy-depleted conventional oil reservoir), this will cause deposition of asphaltic material ahead of the steam front. The ever-present disadvantage is that any deposited material will block the reservoir flow channels, thereby restricting oil flow to the production well. The distinct advantage is that if flow to the recovery well is not restricted, there is the potential for recovery of partially upgraded heavy oil.

Another effect is that the oil bypassed by the advancing condensation front is so volatile that it shrinks significantly as steam continues to pass by, resulting in low recovery of the heavy oil. These effects would be more pronounced with the residual oil from conventional crude oil recovery, which contains relatively high levels of volatile constituents. Overall, the condensation, steam-stripping, and steam-distillation effects result in more favorable recovery than is expected on the basis of viscosity reduction alone.

Steam can also be injected into one or more wells with production coming from other wells (*steam drive*). This technique is very effective in heavy oil formations but has found little success during application to heavy oil reservoirs because of the difficulty in connecting injection and production wells. However, once the flow path has been heated, the steam pressure is cycled, alternately moving steam up into the oil zone, and the oil is allowed to drain down into the heated flow channel to be swept to the production wells.

All of the processes in which hot fluid is injected through the wellhead suffer from heat losses from the injection well bore to the overburden formations. These heat losses can be a significant fraction of the injected heat when the wells are deep or poorly insulated and the injection rates are low. Under such conditions, the temperature of an injected non-condensable fluid entering the formation may be significantly lower than at the wellhead. When the injected fluid is condensable, as in the case of steam, the heat losses cause condensation of the vapor. If the steam is unable to maintain the heat and enters the reservoir as hot-water, the oil in the reservoir actually comes into contact with hot-water rather than steam. Whether or not the reservoir is injected with steam is subject to the depth of the reservoir (i.e., the distance the surface-generated steam has to travel to enter the reservoir). Cooling of the steam by any encroaching water could have a significantly adverse effect not only on the rate of growth of the steam zone but also on the initiation of the steam zone. However, there are several schools of thought that contend that the majority of steam-based processes are, in fact, hot-water-based processes. To overcome the steam-or-hot-water issue, surface lines generally are insulated, and injection well are completed in a manner that reduces heat losses.

7.2.1 Steam Drive Injection (Steam Injection)

Steam drive injection (steam injection) has been commercially applied since the early 1960s. Steam drives differ markedly in performance from hot-water drives, the difference being solely due to the presence and effects of the condensing vapor. The presence of the gas phase causes light components in the crude oil to be distilled and carried along as hydrocarbon components in the gas phase. Where the steam condenses, the condensable hydrocarbon components do likewise, thus reducing the viscosity of the crude oil at the condensation front. Moreover, the condensing steam makes the displacement process more efficient and improves the sweep efficiency. The net effect is that recovery from steam drives is significantly higher than from hot-water drives.

The process occurs in two steps: (1) steam stimulation of production wells (i.e., direct steam stimulation) and (2) steam drive by steam injection to increase production from other wells (i.e., indirect steam stimulation) (Chapter 5).

When there is some natural reservoir energy, steam stimulation normally precedes steam drive. In steam stimulation, heat is applied to the reservoir by the injection of high-quality steam into the production well.

Huff and puff employs a pattern of vertical wells. Typically, pressurized steam is pumped down a well for weeks and sometimes months, thoroughly heating the reservoir near the well. The process is then halted, usually for several weeks; this allows the heavy oil to become separated from the reservoir sand. Then, the heavy oil is artificially lifted from the same well. This technology has been successfully used for many years but is fast being replaced in Canada by the SAGD method. It is still considered applicable in heavy oil regions where SAGD is not suitable.

7.2.2 Cyclic Steam Injection

Cyclic steam injection (Chapter 5) is the alternating injection of steam and production of oil with condensed steam from the same well or wells. This process is predominantly a vertical well process, with each well alternately injecting steam and producing heavy oil and steam condensate. In practice, steam is injected into the formation at greater than fracturing pressure. This is followed by a soak period after which production is commenced. The heat injected warms the heavy oil and lowers its viscosity. A heated zone is created through which the warmed heavy oil can flow back into the well. This is a well-developed process; the major limitation is that less than 30% (usually less than 20%) of the initial oil in place can be recovered.

Cyclic steam stimulation is often the preferred method for production in heavy oil reservoirs that can contain high-pressure steam without fracturing the overburden. The minimum depth for applying cyclic steam stimulation is on the order of 1,000 feet; depth varies depending upon the type and structure of the overlying formations.

Cyclic steam stimulation works best when there are thick pay zones (>10 m) with high porosity sands (>30%). Shale layers that reduce vertical permeability are not a problem for vertical wells that penetrate thick pay zones. However, good horizontal permeability (>1 darcy) is important for production. Recently, cyclic steam stimulation has been applied to wells with multilateral horizontal legs.

There are three phases in cyclic steam stimulation. First, high-temperature, high-pressure steam is injected for up to one month. Second, the formation is allowed to soak for one or two weeks to allow the heat to diffuse and lower the heavy oil viscosity. Third, heavy oil is pumped out of the well until production falls to uneconomic rates, which may take up to one year. Then the cycle is repeated, as many as 15 times, until production can no longer be recovered. Artificial lift is required to bring the heavy oil to surface.

Typical recovery factors for cyclic steam stimulation are 20% to 35% with steam-to-oil ratios (SOR) of 3 to 5.22. Steamfloods may follow cyclic steam stimulation. While cyclic steam stimulation produces the heavy oil around a single well bore, steamflooding recovers the heavy oil between wells. For example, a five-spot pattern with four producing wells surrounding a central steam injection well is a common configuration. The well spacing can be less than two acres for a field in steam flood. The steam heats the oil to lower its viscosity and provides pressure to drive the heavy oil toward the producing wells. In most steamflood operations, all of the wells are steam-stimulated at the beginning of the flood. In a sense, cyclic steam stimulation is always the beginning phase of a steamflood. In some cases, even the steamflood injection wells are put on production for one or two cyclic steam stimulation cycles to help increase initial project production and pay out the high steamflood capital and operating costs.

Cyclic steam stimulation and steamfloods are used in California, western Canada, Indonesia, Oman, and China. California's Kern River production rose from less than 20,000 barrels per day in the late 1950s before cyclic steam stimulation to over 120,000 barrels per day by 1980 after the introduction of cyclic steam stimulation. The Duri field in Indonesia is the world's largest steamflood and produces 230,000 barrels per day with an estimated ultimate recovery factor of 70% in some locations.

Cyclic steam injection is used extensively in heavy-oil reservoirs, tar sand deposits, and in some cases, to improve injectivity prior to steamflooding or in situ combustion operations.

The technique has also been applied to the California tar sand deposits (Bott, 1967) and in some heavy oil reservoirs north of the Orinoco deposits (Franco, 1976; Ballard et al., 1976). The steam-flooding technique has been applied, with some degree of success, to the Utah tar sands (Watts et al., 1982) and has been proposed for the San Miguel (Texas) tar sands (Hertzberg et al., 1983).

Technical challenges for cyclic steam stimulation and steamflooding are primarily related to reducing the cost of steam, which is generated in most locations using natural gas. The economics may be improved by also generating and selling electricity and by using waste heat for co-generation. Alternative fuels (coal, heavy ends, coke) are discussed separately below; they could also reduce the cost of steam generation. Monitoring and controlling the steam front could also reduce costs by redirecting steam to zones where the heavy oil has not been produced. Steam could be shut off from zones that have been successfully swept and directed toward unswept regions.

Gravity override is a natural occurrence in every steamflood. The steam breaks through to the producers, at which time the process turns into a gravity drainage process. The steam chest at the top of the formation expands downward, and the heated heavy oil drains by gravity to the producing wells. Although the geometry configuration is totally different from that of SAGD (described later), the basic physics are the same.

The measurement of the produced fluids (oil, water, and natural gas) at the surface for each well can be used to optimize production by adjusting artificial lift rates and steam injection rates. Downhole fluid-flow measurements could be used to identify which zones are producing oil, water, or gas in a producing well. Monitoring may involve drilling observation wells where permanent sensors may be deployed or where logging can be periodically performed. Downhole temperature and pressure sensors may use fiber-optic or wire line technology. Water and steam saturation outside a observation well's casing can be measured with nuclear spectroscopy logs. Time-lapse, cross-well electrical imaging can be used to identify bypassed heavy oil zones between closely spaced (500 m) observation wells. Cross-well seismic and surface seismic measurements might be used to locate steam fronts. High-resolution imaging of the formation and the fluid saturations before completing the wells and during production, however, is an open technical challenge.

Technologies must be reliable and have long operating periods between service periods. High-temperature- (up to 300°C) and corrosion-resistant equipment including pumps (artificial lift), cements, completions, liners, packers, valves, electronics, and sensors are needed. Thermal expansion of the formation can also cause the casing to fail.

Most cyclic steam stimulation and steamflood wells have been vertical wells. More recently, vertical wells with multilateral branches and horizontal wells are being tried. The advantage is a reduced footprint while tapping large subsurface regions. Optimal control and configuration of these wells for cyclic steam stimulation and steam flood recovery processes are still being developed.

Cyclic steam injection also is used as a precursor to steam drive technology. In reservoirs containing heavy crude oil, the resistance to flow between the wells may be sufficiently high that steam injection rates are severely limited, making steam drive technically inefficient. Cyclic steam injection reduces the flow resistance near wells, where the resistance is most pronounced; this alone improves the injection rate attainable during steam drives by reducing the resistance to flow between wells. Repeated cyclic steam injection reduces the flow resistance still farther from the wells and may lead to connecting the heated zones of adjacent wells and further improving the operability of the steam drive.

A drawback to cyclic steam injection is that the ultimate recovery may be low relative to the total oil in place in the reservoir. Ultimate recoveries from steam drives are generally much larger than those from cyclic steam injection. Thus, cyclic steam injection followed by a steam drive is an attractive combination; crude production is accelerated quickly, and the ultimate recovery is quite high.

7.2.3 Steam Drive

Steam drive involves the injection of steam through an injection well into a reservoir and the production of the mobilized heavy oil and steam condensate from a production well. Steam drive is usually a logical follow-up to cyclic steam injection. Steam drive requires sufficient effective permeability to allow injection of the steam at rates sufficient to raise the reservoir temperature to move the heavy oil to the production well.

Two expected problems inherent in the steam drive process are steam override and reservoir plugging. Any in situ thermal process tends to override (migrate to the top of the effected interval) because of differential density of the hot and cold fluids. These problems can be partially mitigated by rapid injection of steam at the bottom or below the target interval through a high-permeability water zone or fracture. Each of these options will raise the temperature of the entire reservoir by conduction and, to a lesser degree, by convection, and the effectiveness of the following injection of steam into the target interval will be enhanced.

For a successful steam drive project, the porosity of the reservoir rock should be at least 20%, the permeability should be at least 100 millidarcies, and the heavy oil saturation should be at least 40%. The reservoir oil content should be at least 800 bbl per acre-foot. The depth of the reservoir should be less than 3000 feet, and the thickness should be at least 30 feet. Other preferential parameters have also been noted on the basis of success with several heavy oil reservoirs.

7.3 In Situ Combustion Processes

In situ combustion processes are not new; work on various aspects of the processes has continued since at least 1923 (Howard, 1923; Wolcott,1923; Kuhn and Koch, 1953: Grant and Szasz, 1954). In fact, in situ combustion has been applied in over a hundred fields (Farouq Ali, 1972; Chu 1977, 1982; Brigham et al., 1980). The South Belridge project, which began commercial operations in 1964, is of special significance since it was a commercial success and analyses of the field data yielded a number of useful, new concepts and correlations. Certain ideas carried over from waterflooding were found to be inappropriate to the in situ combustion process (Gates and Ramey, 1958, 1980; Ramey et al., 1992).

The main parameters required in the design of an in situ combustion project are, in addition to operating costs, (1) the fuel concentration per unit reservoir volume burned, (2) the composition of the fuel, (3) the amount of air required to burn the fuel, (4) the volume of reservoir swept by the combustion zone, (5) the required air injection rates and pressures, and (6) the oil production rate (Alexander et al., 1962).

In the process, either dry air or air mixed with water can be injected into the reservoir. Ideally, the fire propagates uniformly from the air injection well to the producing well, moving oil and combustion gases ahead of the front. The coke remaining behind the moved oil provides the fuel. Temperatures in the thin combustion zone may reach several hundred degrees centigrade, sufficient to crack the heavy oil into lower-boiling products. The oil is subjected to a combination of miscible displacement by the condensed light hydrocarbons, hot-water drive, vaporization, and steam and gas drive. As the temperature in the volume element exceeds about 345°C (650°F), the oil will more than likely undergo thermal cracking to form a volatile fraction and a low-volatility, coke-like residue. The volatile products are carried in the gas stream, while the coke-like residue is burned as fuel in the combustion zone. The heat generated at the combustion zone is transported ahead of the front by conduction through the formation matrix and by convection of the vapors and liquids (Wu and Fulton, 1971).

The thickness of the combustion zone is variable, and the temperature is usually on the order of 345°C to 650°C (650°–1200°F) but may be difficult to control (this is the issue with many combustion-related recovery processes). As the combustion front moves forward, a zone of clean sand is left behind where only air flows. As a result of distillation and thermal cracking, the quality of the produced oil is improved. For example, in South Belridge, the produced oil gravity was as high as 18° API, compared to 12.9° API for the original oil.

Furthermore, in the South Belridge work, it became evident that reservoir lithology is an important parameter in fuel deposition (Gates and Ramey, 1958), and the amount of fuel deposited increases with the addition of clay to the sample of oil and sand (Bousaid and Ramey, 1968). This is in agreement chemically with the known catalyst activity of minerals in refining processes and the ability of clay to adsorb polar constituents such as the carbenes and carboids that are the thermal precursors to coke (Speight, 2007 and references cited therein).

In situ combustion is normally applied to reservoirs containing low-gravity oil but has been tested over perhaps the widest spectrum of conditions of any enhanced oil recovery process. In the process (Chapter 5), heat is generated within the reservoir by injecting air and burning part of the crude oil. This reduces the oil viscosity and partially vaporizes the oil in place. The oil is driven out of the reservoir by a combination of steam, hot-water, and gas drive.

Injection of air alone is known as dry underground combustion, in situ combustion, or fireflooding (Kuhn and Koch, 1953). However, there are several variants of the in situ combustion process.

In *forward combustion*, the combustion front moves in the same direction as the air flow but *reverse combustion* occurs when the combustion front moves in a direction opposite to the flow of the injected air. Reverse combustion is achieved by igniting the crude near a production well while temporarily injecting air into it. Upon resumption of the normal air injection program, the combustion front will move toward the injection wells.

The fire front can be difficult to control, and it may propagate in a haphazard manner, resulting in premature breakthrough to a producing well. There is a danger of a ruptured well with hot gases escaping to the surface. The produced fluid may contain an oil-water emulsion that is difficult to break, and contrary to expectations, it may also contain heavy-metal compounds that are difficult to remove in the refinery.

The well bore near the pay zone, or for that matter any part of the injection well that might come in contact with free oxygen and fuel (crude oil), should be designed for high thermal stresses. Crude oil is likely to enter the well bore by gravity drainage where the air enters the formation preferentially over a short segment of a large open interval that has adequate vertical permeability. This crude inflow may be increased as the reservoir temperature near the well bore increases as a result of the heat generated either by the ignition system used in the well bore or by the combustion process itself (including reverse combustion following spontaneous ignition a short distance into the reservoir). When designing injection wells, precautions should be taken against any likelihood of combustion in the well bore.

Wet and partially quenched combustion, also known by the acronym COFCAW (*combination of forward combustion and waterflooding*) (Dietz Weijdema, 1968a; Parrish and Craig, 1969), uses water injection during the combustion process to recuperate the heat from the burned zone and adjacent strata. In this process, the ratio of injected water to air is used to control the rate of advance of the combustion front, the size of the steam zone, and the temperature distribution.

The appeal of an in situ combustion process is the potential for partial upgrading of the oil in the reservoir, providing the undesirable constituents of the oil remain in the reservoir. In addition, the process has the rapid kinetics of a thermal process, and there is no need to generate energy at the surface.

During in situ combustion or fireflooding, energy is generated in the formation by igniting heavy oil in the formation and sustaining it in a state of combustion or partial combustion. The high temperatures generated decrease the viscosity of the oil and make it more mobile. Some cracking of the heavy oil occurs, and an upgraded product rather than heavy oil itself is the fluid recovered from the production wells.

The relatively small portion of the oil that remains after the displacement mechanisms have acted becomes the fuel for the in situ combustion process. Production is obtained from wells offsetting the injection locations. In some applications, the efficiency of the total in situ combustion operation can be improved by alternating water and air injection. The injected water tends to improve the utilization of heat by transferring heat from the rock behind the combustion zone to the rock immediately ahead of the combustion zone.

The performance of in situ combustion is predominantly determined by four factors:

1. The quantity of oil that initially resides in the rock to be burned

2. The quantity of air required to burn the portion of the oil that fuels the process

3. The distance to which vigorous combustion can be sustained against heat losses
4. The mobility of the air or combustion product gases

In many field projects, the high gas mobility has limited recovery through its adverse effect on the sweep efficiency of the burning front. Because of the density contrast between air and reservoir liquids, the burning front tends to override the reservoir liquids. To date, combustion has been most effective for the recovery of viscous oils in moderately thick reservoirs in which reservoir dip and continuity provide effective gravity drainage or operational factors permit close well spacing.

The use of combustion to stimulate oil production is regarded as attractive for deep reservoirs. In contrast to steam injection, it usually involves no loss of heat. The duration of the combustion may be less than 30 days or as much as 90 days, depending on requirements. In addition, backflow of the oil through the hot zone must be prevented or coking will occur.

In addition to providing the heat to mobilize the oil, in situ combustion of heavy oil can provide some in situ upgrading through the use of minerals or additives (Dabbous and Fulton, 1972; Fassihi et al., 1984a, 1984b; He, 2004; He et al., 2005; Shallcross et al., 1991; Strycker et al., 1999; Castanier and Kovscek, 2005). During in situ combustion of heavy oils, temperatures of up to 700°C can be observed at the combustion front. This is sufficient to promote some upgrading.

Heavy oil upgrading is of major economic importance. Numerous field observations have shown upgrading of 2° to 6°API for heavy oils undergoing combustion (Ramey et al., 1992).

7.3.1 Forward Combustion

The most common form of in-situ combustion is *dry forward combustion*. In this process, air is injected into a heavy oil reservoir, the crude is ignited in-situ, and the resulting combustion front moves away from the injection well. The heat generated at the combustion front propagates through the reservoir, reduces the oil viscosity, and thereby increases the oil production rate and recovery. The propagation of a combustion front in a reservoir is the most rapid method of thermal recovery. The combustion front can move more rapidly than heat can be moved by conduction and convection in a reservoir, and the convective heat wave velocity for the case of air injection is about one quarter that of the combustion front (Martin et al., 1958; Ramey, 1971).

In forward combustion, the hydrocarbon products released from the zone of combustion move into a relatively cold portion of the formation. Thus, there is a definite upper limit of the viscosity of the liquids that can be recovered by a forward-combustion process. On the other hand, since the air passes through the hot formation before reaching the combustion zone, burning is complete; the formation is left completely cleaned of hydrocarbons.

Generally, forward combustion is referred to as dry forward combustion, and the effects of any reservoir water are, for unknown reasons, ignored. Temperature levels in dry forward combustion, which affect the displacement, distillation, stripping, cracking, and formation of solid fuel downstream of the combustion front, are affected by the amount of fuel burned per unit volume of reservoir rock. At high temperatures (approximately 815°C, 1500°F), the combustion zone is very thin, whereas at lower temperatures (approximately 345°C, 650°F), a smoldering reaction with the bypassed air may occur over distances of several feet and generate heat and ultimately cause spontaneous ignition. At intermediate temperatures (approximately 600°C, 1200°F), the combustion reaction proceeds slowly enough to allow significant leakage of free oxygen in the direction of flow, thus increasing the thickness of the reaction zones.

A characteristic of the dry forward combustion process is that the temperature of the burned zone remains quite high because the heat capacity of the air injected is too low to transfer a significant amount of heat. For this reason, water sometimes is used during or after the combustion process to help transfer the heat from the burned zone to downstream areas.

Another form of in situ combustion is the *wet combustion* method, in which air and water are injected concurrently or alternately. The purpose of injecting water is to recuperate and transport heat from the burned zone to the colder regions downstream of the combustion front. This method may be considered for thin reservoirs, where heat loss to adjacent formations is significant (Dietz and Weijdema, 1963; Parrish and Craig, 1969; Dietz, 1970; Beckers and Harmsen, 1970; Burger and Sahuquet, 1973).

The addition of water during the combustion process means that heat is transferred more effectively than with air alone. In addition, the steam zone ahead of the combustion front is larger, and the reservoir is swept more efficiently than with air alone. The improved displacement from the steam zone results in lower fuel availability and consumption in the combustion zone, so a greater volume of the reservoir is burned for a given volume of air injected.

Water must be injected in the wet combustion process, but it may be difficult to inject both air and water simultaneously at the desired rates in low-permeability reservoirs. In such cases, the water and air can be injected alternately, and the duration of the air and water injection periods can be controlled to achieve the desired average water/air ratio, which is essential to obtain the desired combustion-front velocity and temperature level. At a low water/air ratio, the water that reaches the combustion front has been converted to steam, whereas at a high water/air ratio, the water that reaches the combustion front is, for the most part, in the liquid phase.

One of the benefits of the combustion process is the production of a partially upgraded product. The temperature gradient ahead of the combustion front either causes the lower-molecular-weight (more volatile) constituents to distill and move toward the cooler portion of the reservoir and mix with unheated oil or some of the higher-molecular-weight (less volatile) constituents crack (thermally decompose) and the volatile products also moves toward the cooler portion of the reservoir and mix with unheated oil. The cracking process also produces a carbonaceous residue that deposits on the reservoir rock and is consumed as fuel during the combustion process.

7.3.2 Reverse Combustion

A third variation of the in situ combustion process is the *reverse combustion* method. In this technique, the combustion zone is initiated at a production well. The reverse combustion front travels countercurrent to the air towards the injection well where air is injected. The oil flows towards the production well, through the combustion zone. Since no oil bank is formed, the total flow resistance decreases with time, and thus this method is particularly suitable for reservoirs containing very viscous crude oils. One disadvantage of this method is the likelihood of spontaneous ignition. Spontaneous ignition would result in oxygen being consumed near the injector, and the process would change to forward combustion (Dietz and Weijdema, 1968b).

Another disadvantage of reverse combustion is the inherent instability of the process, which results in narrow combustion channels being formed and therefore an inefficient burn (Gunn and Krantz, 1980; Johnson et al., 1980).

Reverse combustion is particularly applicable to reservoirs with lower effective permeability, in contrast with forward combustion. The method is more effective because the lower permeability would cause the reservoir to be plugged by the mobilized fluids ahead of a forward combustion front. In the reverse combustion process, the vaporized and mobilized fluids move through the heated portion of the reservoir behind the combustion front. The reverse combustion partially cracks the oil, consumes a portion of the oil as fuel, and deposits residual coke on the sand grains, leaving 40% to 60% of the oil as recoverable oil. This coke deposition serves as a cementing material, reducing movement and production of sand.

As the combustion front reaches the heavy oil, a significant amount of cracking occurs, and a relatively large amount of solid fuel (compared to the amount formed during forward combustion) is deposited on the reservoir rock. However, recovery of the partially upgraded product is lower than in forward combustion because some is burned in the process, and a high degree of *equivalent oil saturation* occurs in the burned zone (including unburned solid fuel). On the other hand, the API gravity of the recovered product is increased significantly by extensive cracking because the products flow through the hot burned zone and are subject to secondary, tertiary, or higher level cracking reactions (Figure 7–3).

Historically, reverse combustion has been difficult to maintain because the oxygen is depleted not far from the injection well (Dietz and Weijdema, 1968b). Furthermore, sustained air injection into an unheated reservoir generally leads to spontaneous ignition near the injection well (Elkins et al., 1974; Burger, 1976; Tadema and Weijdema, 1970).

The addition of water or steam to an in situ combustion process can result in a significant increase in the overall efficiency of the process. Two major benefits may be derived. Heat transfer in the reservoir is improved because the steam and condensate have greater heat-carrying capacity than combustion gases and gaseous hydrocarbons. Sweep efficiency may also be improved because of the more favorable mobility ratio of steam-oil compared with gas-oil.

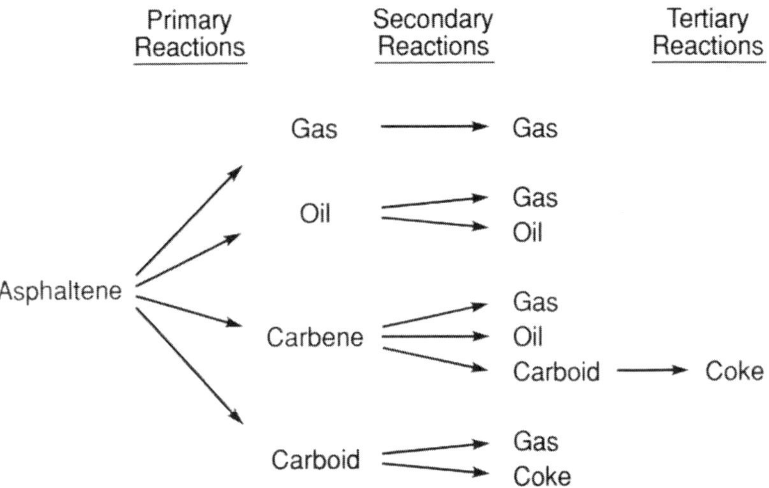

Figure 7–3 *Multilevel cracking reactions using the asphaltene constituent as an example.*

Process efficiency is affected by reservoir heterogeneity that reduces horizontal sweep. The underburden and overburden must provide effective seals to avoid loss of injected air and produced oil. Process efficiency is enhanced by the presence of some interstitial water saturation. The water is vaporized by the combustion and enhances the heat transfer by convection. The combustion processes are subject to override because of differences in the densities of injected and reservoir fluids. Production wells should be monitored for and equipped to cool excessively high temperatures (>1095°C, >2000°F) that may damage downhole production tools and tubulars.

Applying a preheating phase before the recovery phase may significantly enhance the steam or combustion extraction processes. Preheating can be particularly beneficial if the saturation of the more viscous oil (API gravity = 12° or lower) is sufficiently great to lower the effective permeability to the point that production is precluded by reservoir plugging. Preheating increases the mobility of the oil by raising its temperature and lowering its viscosity, and the outcome is a lower required pressure to inject steam or air to recover the oil.

Using combustion to stimulate oil production is regarded as attractive for deep reservoirs. In contrast to steam injection, it usually involves no loss of heat. The duration of the combustion may be less than 30 days or approximately 90 days, depending upon requirements. In addition, backflow of the oil through the hot zone must be prevented or coking occurs.

When it is determined that the reservoir should be preheated, there are several methods by which this can be accomplished. Conducting a reverse combustion phase in a zone of relatively high effective permeability and low oil saturation is one method. Steam or hot gases may he rapidly injected into a high-permeability zone in the lower portion of the reservoir. In the *fracture-assisted steam technology (FAST)* process, steam is injected rapidly into an induced horizontal fracture near the bottom of the reservoir to preheat the reservoir. This process has been applied successfully in three pilot projects in southwest Texas. Shell has accomplished the same preheating goal by injecting steam into a high-permeability bottom water zone in the Peace River (Alberta) field. Electrical heating of the reservoir by radio-frequency waves may also be an effective method.

A variation of the combustion process involves use of a heat-up phase, a blow-down (production) phase, and then a displacement phase using a fire-water flood (a combination of forward combustion and waterflood, COFCAW). This modified combustion approach has been applied to the Athabasca deposit. In this manner, over an 18-month period (heat-up: 8 months; blow-down: 4 months; displacement: 6 months), 29,000 bbl of upgraded oil were produced from an estimated 90,000 bbl of oil in place.

In any field in which primary recovery operations are followed by secondary or enhanced recovery operations, there is a change in product quality. Product oils recovered by the thermal stimulation of heavy oil reservoirs show some improvement in properties over those of the heavy oil in place. Although this improvement in properties may not appear to be too drastic, nevertheless it usually is sufficient to have major advantages for refinery operators. Any incremental increase in the units of the hydrogen/carbon ratio can save amounts of costly hydrogen during upgrading. The same principles are also operative for reductions in the nitrogen, sulfur, and metals content. This removal of nitrogen, sulfur, and metals from the products also improves catalyst life and activity when the product oil is refined. In short, in situ recovery processes may have the added benefit of *leaving* some of the more obnoxious constituents (from the processing objective) in the ground.

A combustion displacement process may be more attractive than a steam drive process, assuming that a downhole steam generator is also not available to the reservoir. Conditions that might preclude a steam drive and be in favor of a combustion process include: (1) high sustained injection pressures—above 1,500 psi; (2) excessive heat losses from the injection well in reservoirs more than 4,000 feet deep; (3) a lack of a supply of fresh water or treatment costs that make the use of steam prohibitively expensive; (4) serious clay swelling problems; and (5) thin or low-porosity sands where wet combustion processes tend to be more efficient at heat management than steam drives (Wilson and Root, 1966).

Wet combustion would be considered instead of dry combustion where there is ample available water and where water/air injectivity is favorable. Wet combustion would not be used where there is little likelihood that the water would move through the burned zone to recuperate heat effectively, as in gravity-dominated operations (Koch, 1956; Gates and Sklar, 1971).

7.3.3 The THAI Process

The *toe-heel-air-injection (THAI) process*, although a combustion process, deserves special mention in a separate section because it offers a way to control the forefront and produce a product that is partially upgraded and ready for the refinery. The THAI process is a new combustion process that combines a vertical air injection well with a horizontal production well (Figure 7–4). During the process, a combustion front is created where part of the oil in the reservoir is burned, generating heat that reduces the viscosity of the oil, allowing it to flow by gravity to the horizontal production well. The combustion front sweeps the oil from the toe to the heel of the horizontal producing well, recovering an estimated 80% of the original oil in place while partially upgrading the crude oil in situ.

The THAI process has potential to operate in reservoirs that are lower in pressure, of a lower quality, thinner, and deeper than required for the SAGD process. In addition, the THAI process integrates existing proven technologies and provides the opportunity to create a step change in the development of heavy oil resources globally.

The horizontal well arrangement provides a unique gravity and pressure drawdown geometry. Another key feature is that oil recovery occurs via a short displacement mechanism, which requires oil to

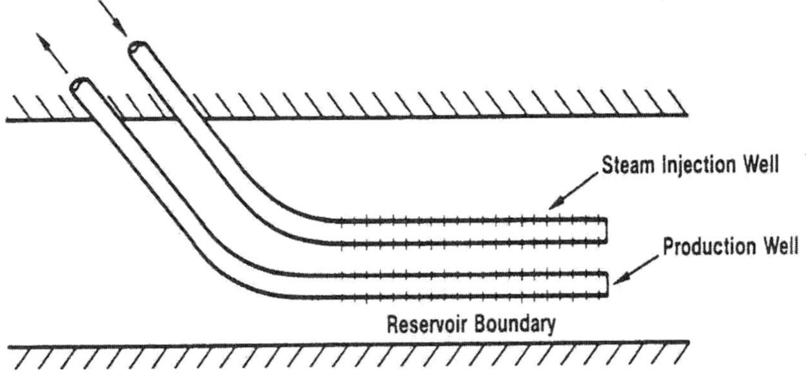

Figure 7–4 *Steam assisted gravity drainage (SAGD).*

move downwards (with the help of gravity) typically just 15 to 30 feet, as opposed to the lateral movement of several hundred feet in the usual combustion processes.

In comparison to SAGD, fewer surface facilities are required. They are mainly an electric powered air compressor for injection and separators and tanks for the production well.

The THAI process begins with preheating both well bores using steam to initiate oil mobility and clear pore space between the injector and the toe of the producing well. After ignition (auto-ignition with some oils), the energy to sustain in situ combustion comes from the burning of the coke that is continuously laid down within the reservoir. Product sulfur is reduced, as are heavy metals, which are left as an inert residue on the reservoir rock. No water or gas fuel is required during production, and the produced water can be treated to usable industrial quality.

A further benefit of the THAI process is that it performs in situ upgrading through thermal cracking of the heavy oil. Laboratory tests achieved upgrading by up to 10°API. On this basis, a 10° API oil could be expected to yield an 18° to 20° API oil at the surface. This is a very desirable feature of any recovery process since every increase of 1° API can mean refinery savings of several dollars per barrel.

The reservoir matrix is an important aspect of the THAI process; some clay content is required to help thermal cracking of the oil ahead of the combustion front. The process can break down intermittent shale breaks that could otherwise obstruct permeability. The THAI process can also cope with a bottom water zone by effectively steaming it out. Tests also indicate that the THAI process can handle a gas cap; gravity and pressure draws down all the gasses ahead of the combustion front.

Ahead of the combustion front (typically around 600°C, 1112°F) is the coking zone, in front of which is a 10-to-15-feet-wide mobile oil zone through which drainage takes place into the horizontal well. These zones move through the reservoir at about one to three feet per day, depending on the air injection rate. The temperature drops to between 200°C and 350°C at the front of the mobile oil zone, with a corresponding reduction in the rate of drainage. Ahead of the mobile oil zone is the cold immobile virgin oil layer, through which there is no communication for gas. This characteristic of the process geometry means that the only way out is down into the open section of the horizontal well. The horizontal well trajectory is thus a built-in self-controlling guidance system for fluid flow. This makes the process much more controllable than conventional in situ recovery systems, in which fluids are less controlled and can move and penetrate anywhere in the reservoir.

In summary, the THAI process has many potential technical benefits including (1) higher source recovery that is estimated at 70% to 80% of the oil in place, and it is potentially feasible for use over a broader range of reservoirs, including reservoirs having low pressure, thin reservoirs, and previously steamed reservoirs; (2) well geometry that enforces a short flow path so that the instabilities associated with the longer flow path in conventional combustion methods are reduced or even eliminated; and (3) a lower environmental impact insofar as there is negligible fresh water use, less greenhouse gas emissions, a smaller surface footprint, and easier reclamation.

As an extension of the THAI process, the CAPRI process involves a layer of refinery-type catalyst along the outside of the horizontal producer well; it is therefore the catalytic variant of the THAI process. It uses an annular sheath of solid catalyst surrounding the horizontal producer well in the bottom of the oil layer. The thermally cracked oil produced drains into the horizontal producer well, first passing through the layer of catalyst where the high pressure and temperature in the reservoir enable thermal cracking and hydroconversion reactions to take place so that only light, converted oil is produced at the surface.

7.4 Other Processes

Many innovative concepts in heavy oil production have been developed. The major new technologies that have positively affected the heavy oil industry in the last 10 years are:

1. Horizontal well technology for shallow applications (<1000 m), often combined with gravity drainage approaches.
2. Cold production using long horizontal wells that are often combined with multilaterals with aggregate lengths exceeding several kilometers.
3. Gravity-driven processes, particularly steam-assisted gravity drainage, vapor-assisted petroleum extraction, and inert gas injection, all using horizontal wells to establish stable, gravity-assisted oil recovery.
4. Cold heavy oil production with sand, or CHOPS technology, where sand production is encouraged and managed as a means of enhancing well productivity.
5. Pressure pulse flow enhancement technology, both as a reservoirwide method and as a workover method.
6. Improvements in upgrading viscous, sulfur-rich heavy crude oil feedstocks.

The oil industry pioneered drilling shallow (500 to 3500 feet deep) horizontal wells at cost-per-meter values that are now only 1.2 to 1.3 times those of vertical wells. These wells, in the shallowest cases (150 to 1500 feet deep) are often drilled using masts inclined to reduce curvature build rates required to *turn the corner* from vertical to horizontal. In *reduced footprint* developments, where there are a number of producing wells and injecting wells on the same pad (savings in roads, services, etc.), many horizontal wells (2 to 6) may be drilled from a small area, no larger than a hectare.

7.4.1 Horizontal Well Technology

Coiled tubing drilling and workover have been introduced and perfected in the last decade, further reducing costs of horizontal well drilling. Good seismic control and cuttings analysis allows precise steering in thin zones (<20 ft) to place the well in the optimum position in the reservoir. In the production phase, the long drainage length of a single well, as much as 1200 m in many cases, allows much more effective production. It also gives higher production percentages of original oil in place when used in gravity drainage technologies.

Long horizontal wells with several multilateral branches have been used widely in the development of the heavy oils of Venezuela, where production rates as high as 2000 to 2500 bbl/day in some wells has been achieved through the use of aggregate horizontal lengths as large as 10,000 meters in oil of 1200 to 5000 centipoises viscosity. Unfortunately, this technology can only achieve 8% to 15% recovery and only from the best high-permeability zones. Thus, for the more efficient development of these resources, other technologies will necessarily be implemented in the future.

Horizontal wells using cold primary production of heavy oils have been widely used in Canada since the late 1980s, but the successes (usually well published) have been substantially offset by the failures (rarely published). Furthermore, low recovery factors (seldom close to 10%), early water breakthrough (usually impossible to plug when it happens), short well life (30% to 40% decline per year), and other factors, such as expensive workover if sand plugging takes place, have combined to make this a technology that has little attraction in Canada.

7.4.2 Inert Gas Technology

Inert gas injection (IGI) is a technology for conventional oils in reservoirs where good vertical permeability exists or where it can be created through propped hydraulic fracturing. It is generally viewed as a *top-down* process with nitrogen or methane injection through vertical wells at the top of the reservoirs, which creates a gas-oil interface that is slowly displaced toward long horizontal production wells. As with all gravity drainage processes, it is essential to balance the injection and production volumes precisely so that the system does not become pressure driven but rather remains in the gravity-dominated flow regime (Meszaros et al., 1990).

High recovery ratios are achieved because, in the absence of elevated pressure gradients, a thin oil film is maintained between the gas and water phases. The film is maintained because the sum of the oil-water and gas-oil surface tensions is always less than the water-gas surface tension. Thus, the thin film configuration is thermodynamically stable, and it allows the oil to drain to values far lower than the *residual oil saturation value.*

The interfaces in IGI are gravity-stabilized because of the difference in phase densities; at slow drainage rates, the interfaces remain approximately horizontal without viscous fingering. In one configuration, the horizontal wells are produced under a back-pressure equal to the pressure in the underlying water phase, so no water coning can occur. During production, if the water cut increases, the production rate is reduced so that the interface becomes stable. Alternatively, if gas is injected too quickly, gas coning can develop. If this is observed, the gas injection rate must be reduced to sustain stability. The process is continued until the oil zone is "pinched down" to the horizontal well, achieving the high recovery ratios possible with gravity drainage methods. As with all gravity drainage processes, it is generally necessary to place the horizontal wells as low in the structure as possible. These principles are fundamental to all gravity-dominated processes, and failure to adhere to them will drive the system into conditions of instability (such as coning and fingering).

In reservoirs with excellent vertical permeability, the bottom water zone can also be injected with water to cause the oil-water interface to rise slowly toward the production well. First implemented in Canada, IGI is used extensively in carbonate pinnacle reefs with excellent vertical permeability. Recovery ratios exceeding 80% are systematically achieved.

7.4.3 Steam-Assisted Gravity Drainage (SAGD)

Steam-assisted gravity drainage (SAGD) was initially developed to recover bitumen from the Canadian oil sands (Dusseault et al., 1998). The key element of SAGD is that the two wells need to be parallel and horizontal. It is only in the last 10 to 15 years that directional drilling technology has been able to achieve these two characteristics with any degree of certainty.

This method involves drilling two parallel horizontal wells, one above the other, along the reservoir itself (Figure 7–5). The top well is used to introduce hot steam into the oil sands. As the heavy oil thins and separates from the sand, gravity causes it to drain into the lower well, from where it is pumped to the surface for processing. Even though the injection and production wells can be very close (between

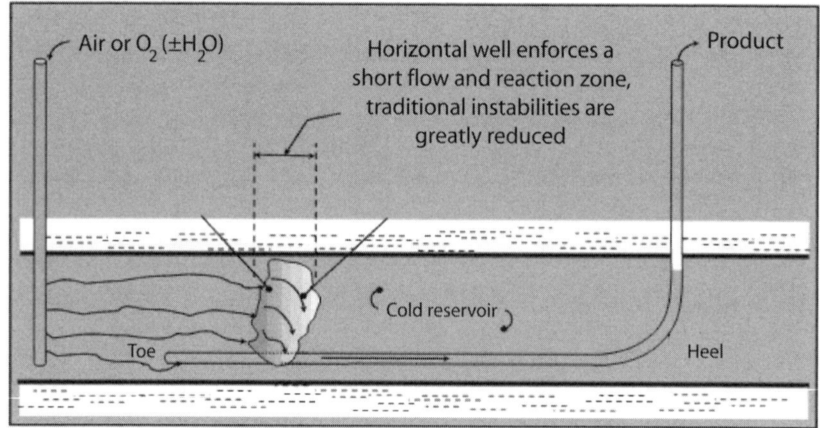

Figure 7–5 *The THAI process.*

5 and 7 m), the mechanism causes the steam-saturated zone (known as the steam chamber) to rise to the top of the reservoir, expand gradually sideways, and eventually allow drainage from a very large area. The method is claimed to significantly improve heavy oil recovery by between 50% and 60% of the original oil in place (OOIP) and is therefore more efficient than most other thermal recovery methods.

The key benefits of the SAGD process are an improved steam-oil ratio and high ultimate recovery (on the order of 60% to 70%). The outstanding technical issues relate to low initial oil rate, artificial lifting of bitumen to the surface, horizontal well operation, and the extrapolation of the process to reservoirs having low permeability, low pressure, or bottom water.

In the process, a pair of horizontal wells, separated vertically by about 15 to 20 feet are drilled at the bottom of a thick unconsolidated sandstone reservoir. Steam, perhaps along with a mixture of hydrocarbons that dissolve into the oil and help reduce its viscosity, is injected into the upper well. The heat reduces the oil viscosity to values as low as 1 to 10 centipoises (depending on temperature and initial conditions) and develops a steam chamber that grows vertically and laterally. The steam and gases rise because of their low density, and the oil and condensed water are removed through the lower well. The gases produced during SAGD tend to be methane, with some carbon dioxide and traces of hydrogen sulfide.

To a small degree, the noncondensable gases tend to remain high in the structure, filling the void space and even acting as a partial "insulating blanket" that helps to reduce vertical heat losses as the chamber grows laterally. At the pore scale, and at larger scales as well, flow is through countercurrent, gravity-driven flow, and a thin and continuous oil film is sustained, giving high recoveries estimated to be as large as 70% to 80% in suitable reservoirs.

Operating the production and injection wells at approximately the same pressure as the reservoir eliminates viscous fingering and coning processes and also suppresses water influx or oil loss through permeable streaks. This keeps the steam chamber interface relatively sharp, and reduces heat losses considerably. Injection pressures are much lower than the fracture gradient, which means that the chances of breaking into a thief zone (an instability problem that plagues all high-pressure steam injection processes, such as cyclic steam soak), are essentially zero.

The SAGD process, like all gravity-driven processes, is extremely stable because the process zone grows only by gravity segregation, and there are no pressure-driven instabilities such as channeling, coning, or fracturing. It is vital in the SAGD process to maintain a volume balance, replacing each unit volume withdrawn with a unit volume injected to maintain the processes in the gravity-dominated domain. If bottom-water influx develops, this indicates that the pressure in the water is larger than the pressure in the steam chamber, and steps must be taken to balance the pressures. Because it is not possible to reduce the pressure in the water zone, the pressure in the steam chamber and production well region must be increased. This can be achieved by increasing the operating pressure of the steam chamber through the injection rate of steam or through reduction of the production rate from the lower well. After some time, the pressures become more balanced and the water influx ceases. Thereafter, maintaining the volume balance carefully is necessary.

Clearly, a low pressure gradient between the bottom water and the production well must be sustained. If pressure starts to build up in the steam chamber zone, then loss of hot water can take place as well. In such cases, the steam chamber pressure must be reduced and perhaps the production rate must be increased slightly to balance the pressures. In all these cases, the system tends to return to a stable configuration because of the density differences between the phases.

SAGD seems to be relatively insensitive to shale streaks and similar horizontal barriers, even up to several meters thick (3 to 6 feet), that otherwise would restrict vertical flow rates. This occurs because as the rock is heated, differential thermal expansion causes the shale to be placed under a tensile stress, and vertical fractures are created, which serve as conduits for steam (up) and liquids (down). As high temperatures hit the shale, the kinetic energy in the water increases, and adsorbed water on clay particles is liberated. Thus, instead of thermal expansion, dehydration (loss of water) occurs, and this leads to volumetric shrinkage of the shale barriers. As the shale shrink, the lateral stress (fracture gradient) drops until the pore pressure exceeds the lateral stress, which causes vertical fractures to open. The combined processes of gravity segregation and shale thermal fracturing make SAGD so efficient that recovery ratios of 60% to 70% are probably achievable even in cases where there are many thin shale streaks. However, there are limits on the thickness of the shale bed that can be traversed in a reasonable time.

Heat losses and deceleration of lateral growth mean that there is an economic limit to the lateral growth of the steam chamber. This limit is thought to be a chamber width of four times (4x) the vertical zone thickness. For thinner zones, horizontal well pairs would therefore have to be placed close together, increasing costs as well as providing lower total resources per well pair. Consequently, the zone thickness limit (net pay thickness) must be defined for all reservoirs.

The cost of heat is a major economic constraint on all thermal processes. Currently, steam is generated with natural gas, and when the cost of natural gas rises, operating costs rise considerably. SAGD is about twice as thermally efficient as cyclic steam stimulation for similar cases, with steam-oil ratios that are now approaching two (instead of four for cyclic steam soak). If high recovery ratios are possible, SAGD will likely displace pressure-driven thermal processes in all cases where the reservoir is reasonably thick and favorable to the process; not all reservoirs are favorable to the SAGD process. Additionally, because of the lower pressures associated with SAGD in comparison to high-pressure processes such as cyclic steam soak and steam drive, greater well bore stability reduces substantially the number of sheared wells that are common in cyclic steam soak projects.

SAGD is not suitable for all reservoirs. It performs best in clean, continuous sands, and it requires continuous vertical permeability. It did not work well on the Clearwater formation at Cold Lake (Alberta,

Canada), for example, in part because the Clearwater formation is neither as permeable nor as vertically continuous as the McMurray (Hart Energy, 2006).

A new concept related to the VAPEX process but used in a different manner can be used in conjunction with a steam-based process, such as SAGD. In the N-Solv process (http://www.n-solv.com/process.htm) (Canadian Patent Applications, 2299790, 2351148, 2374115), heated solvent vapor is injected at moderate pressures into the gravity drainage chamber. The vapor flows from the injection well to the colder perimeter of the chamber, where it condenses. This delivers heat and fresh solvent directly to the bitumen extraction interface. The extraction conditions are mild compared to in situ steam processes, so the valuable components in the bitumen are preferentially extracted, and the problematic high-molecular-weight, coke-forming species (asphaltenes) are left behind. The condensed solvent and oil then drain by gravity to the bottom of the drainage chamber and are recovered via the production well.

7.4.4 Hybrid Processes

Hybrid approaches that involve the simultaneous use of several technologies are evolving and will see greater applications in the future. Some of these options that will be tried at the field scale in the next decade include:

1. A mixture of steam and miscible and noncondensable hydrocarbons is being field tested as a hybrid SAGD-VAPEX approach, with apparent good success and reduction of steam-oil ratios.

2. Single horizontal laterally offset wells can be operated as moderate-pressure cyclic steam stimulation wells in combination with SAGD pairs to widen the steam chamber and reduce steam-oil ratios by about 20%.

3. Simultaneous CHOPS and SAGD, with CHOPS used in offset wells until steam breakthrough occurs and oil recovery commences. Then the CHOPS wells are converted to slow gas and hot-water (or steam) injection wells to control the process. The high-permeability zones generated by CHOPS should accelerate the SAGD recovery process.

4. Incorporating PPT along with CHOPS has already been field tested with economic success, and PPT has potential applications in other hybrid approaches.

5. PPT may aid in partially stabilizing waterflood by reducing the viscous fingering and coning intensity.

In addition to hybrid approaches, the new production technologies and older, pressure-driven technologies will be used in successive phases to extract more oil from reservoirs, even from reservoirs that have been abandoned after primary exploitation. Old reservoirs can be redeveloped with horizontal wells; the wells can even be linked up to bypassed oil because of the physics of oil film spreading between water and gas phases. These staged approaches hold the promise of significantly increasing recoverable reserves worldwide, not just in heavy oil cases.

7.5 In Situ Upgrading

Finally, recent developments in *upgrading* of heavy oil (Ancheyta and Speight, 2007; Speight, 2007) indicate that the near future could see a reduction of the differential cost of upgrading heavy oil. These processes are based on a better understanding of asphaltene solubility effects at high temperatures, the incorporation of a catalyst that is chemically precipitated internally during the upgrading, and the improvement of hydrogen addition or carbon rejection.

In situ upgrading can reduce the viscosity of heavy oil by cracking long hydrocarbon chains and can improve oil quality by reducing or removing asphaltenes and resins. Asphaltenes may contain iron, nickel, and vanadium, which are damaging to refineries. Excess carbon, in the form of coke, may be left in the reservoir.

The upgraded oil flows more readily into the well bore (increasing recovery factor), is easier to lift to surface, and may eliminate the need for a diluent for pipeline transportation. Furthermore, in situ upgrading might eliminate the need for surface upgrading facilities, thus reducing capital investments. In a conventional thermal process (e.g., SAGD), the heavy oil is heated in situ, but it may cool after being produced to surface. It then has to be reheated for upgrading. Therefore, in situ upgrading may be more energy efficient as well.

There are three main approaches for heating the reservoir: steam injection, in situ combustion, and electric heating. Steam injection pressures are limited because most heavy oil deposits are relatively shallow. The maximum steam temperature is limited by the ideal gas law. For example, at a 1,000 m depth, the formation pressure is approximately 10 MPa, which permits a steam temperature of approximately only 300°C. This is too low to provide significant upgrading on a short time scale. In situ combustion is capable of much higher temperatures (approximately 700°C), which should allow significant upgrading. Electric heating (resistance, induction, or radio frequency) should also be able to achieve the high temperatures required for in situ upgrading (Mut, 2005).

Heavy oil constituents can be cracked into lighter hydrocarbon molecules at high-enough temperatures and pressures. In the pyrolysis of heavy oil, carbon-carbon bonds in the hydrocarbon chain are broken by heat; essentially the vibrational energy exceeds the chemical energy in the carbon-carbon bonds. Pyrolysis occurs in the absence of oxygen or a catalyst, but steam may be present. For example, steam cracking and thermal cracking are done in refineries at temperatures at or above 800°C. Such high temperatures are difficult to achieve in the reservoir. Pyrolysis can still occur at lower temperatures, but at much, much slower rates. For example, heavy oil produced under primary and fireflood conditions showed a gradual increase in density, viscosity, and other properties over time.

Adding a catalyst (such as iron) to a thermal process may enhance in situ upgrading, even at the lower temperatures for steam injection (Jiang et al., 2005). Laboratory experiments combining in situ combustion with a catalyst in a horizontal producing well produced significantly upgraded oil. Thermal cracking occurred in the combustion zone, and additional upgrading was achieved by catalytic cracking in the production well (Xia et al., 2002). The downhole catalytic upgrading produced light oil, characterized by a low viscosity, that was readily converted into gasoline and diesel fractions, with a higher conversion in a refinery fluid catalytic cracking (FCC) unit than that obtained with normal virgin bitumen vacuum gas oil (Greaves and Xia, 2004).

7.6 References

Alexander, J.D., Martin, W.L., and Dew, J.N. 1962. Factors Affecting Fuel Availability and Composition During In-Situ Combustion. Journal of Petroleum Technology. 4:1154–1164.

Ancheyta, J., and Speight, J.G. 2007. Hydroprocessing of Heavy Oils and Residua. CRC-Taylor & Francis Group, Boca Raton, Florida.

Advanced Resources International 2005. Basin Oriented Strategies for CO_2 Enhanced Oil Recovery: Onshore California Oil Basins. Prepared for the US Department of Energy, Washington, DC.

Ballard, J.R., Lanfranchi, E.E., and Vanags, P.A. 1976. Proceedings. Twenty-seventh Annual Meeting Petroleum Society, Canadian Institute of Mining. Calgary, Alberta, Canada, June.

Beckers, H.L., and Harmsen, G.J. 1970. The Effect of Water Injection on Sustained Combustion in a Porous Medium. Society of Petroleum Engineers Journal. June:145–163.

Bott, R.C. 1967. Journal of Petroleum Technology. 19:585.

Bousaid, I.S., and Ramey, H.J. Jr. 1968. Oxidation of Crude Oil in Porous Media. Society of Petroleum Engineers. Journal. June:137–148.

Brigham, W.E., Satman, A., and Solitaire, M.Y. 1980. Recovery Correlations for In-Situ Combustion Field Projects and Application to Combustion Pilots. Journal of Petroleum Technology. December:2132–2138.

Burger, J.G. and Sahuquet, B.C. 1973. Laboratory Research on Wet Combustion. Journal of Petroleum Technology. October:1137–1146.

Burger, J.G. 1976. Spontaneous Ignition in Oil Reservoirs. Society of Petroleum Engineers Journal. April:73–81.

Burger, J.G. 1978. Developments in Petroleum Science, No. 7, Bitumens, Asphalts and Tar Sands, ed. G.V. Chilingarian and T.F. Yen, 191. Elsevier, New York.

Craig, F.F. Jr. 1971. The Reservoir Engineering Aspects of Waterflooding. Society of Petroleum Engineers, Dallas, Texas.

Castanier, L. M., and Kovscek, A. R. 2005. Heavy-Oil Upgrading In Situ via Solvent Injection and Combustion: A "New" Method. European Association of Geoscientists and Engineers 67th Conference and Exhibition. Madrid, Spain, June 13–16.

Chu. C. 1977. A Study of Fireflood Field Projects. Journal of Petroleum Technology. February:171–179.

Chu, C. 1982. State-of-the-Art Review of FireFlood Field Projects. Journal of Petroleum Technology. January:19–36.

Curtis, C., Kopper, R., Decoster, E., Guzmán-Garcia, A., Huggins, C., Knauer, L., Minner, M., Kupsch, N., Linares, L.M., Rough, H., and Waite, M. 2002. Heavy Oil Reservoirs. Oilfield Review. Autumn:30–51.

Dabbous, M.K., and Fulton, P.F. 1972. Low Temperature Oxidation Kinetics and Effects on the In-Situ Combustion Process. Paper No. SPE 4143. Society of Petroleum Engineers-American Institute of Mechanical Engineers 47th Annual Fall Meeting. San Antonio, Texas, Oct. 8–11.

Dietz, D.N., and Weijdema, J. 1963. Wet and Partially Quenched Combustion. Journal of Petroleum Technology. April:411–415.

Dietz, D.N. 1967. hot-water Drive. Proceedings. Volume 3, 451–457. Seventh World Petroleum Congress. Mexico City, Mexico.

Dietz, D.N., and Weijdema, J. 1968a. Wet and Partially Quenched Combustion. Journal of Petroleum Technology. 20:411–413.

Dietz, D.N., and Weijdema, J. 1968b. Reverse Combustion Seldom Feasible. Producers Monthly. 32(5):10.

Dietz, D.N. 1970. Wet Underground Combustion, State of the Art, Journal of Petroleum Technology. May:605–617.

Dusseault, M.B., Geilikman M.B., and Spanos T.J.T. 1998. Journal of Petroleum Technology. 50(9):92–94.

Elkins, L.F., Skov, A.M., Martin, P.J., and Lutton, D.R. 1974. Experimental Fireflood—Carlyle Field, Kansas. Paper No. SPE 5014. Society of Petroleum Engineers Annual Meeting, Houston, Texas, October 6–9.

EPRI. 1999. Enhanced Oil Recovery Scoping Study. Report TR-113836. Electric Power Research Institute, Palo Alto, California.

Farouq Ali, S.M. 1972. A Current Appraisal of In-Situ Combustion Field Tests. Journal of Petroleum Technology. April:477–486.

Fassihi, M.R., Brigham, W.E., and Ramey, H.H. 1984a Reaction Kinetics of In-Situ Combustion: Part 1– Observations. Society of Petroleum Engineers Journal. August:399–416.

Fassihi, M. R., and W. Brigham. 1984b.Reaction Kinetics of In-Situ Combustion. Society of Petroleum Engineers Journal (Sept):399–416.

Franco, A. 1976. Oil and Gas Journal. 74(14):132.

Gates, G.F., and Ramey, H.J. Jr. 1958 Field Results of South Belridge Thermal Recovery Experiment. Transactions American Institute of Mechanical Engineers. 213:236–244.

Gates, G.F., and Ramey, H.J. Jr. 1980. Method of Engineering In-Situ Combustion Oil-Recovery Projects. Journal of Petroleum Technology. February:285–294.

Gates, C.F., and Sklar, I. 1971. Combustion as a Primary Recovery Process—Midway Sunset Field. Journal of Petroleum Technology. Pet. 23:981–986.

Grant, B.F., and Szasz, S.E. 1954. Development of an Underground Heat Wave for Oil Recovery. Journal of Petroleum Technology. May:22–23.

Greaves, M., and Xia, T. X. 2004. Journal of Canadian Petroleum Technology. 43(9):25–30.

Grim, R.E. 1968. Clay Mineralogy, McGraw Hill: New York.

Gunn, R.D., and Krantz, W.B. 1980. Reverse Combustion Instabilities in Tar Sands and Coal, Society of Petroleum Engineers Journal. August:267–277.

Hart Energy. 2006. Heavy Oil: Unleashing the Potential. Supplement to E&P Oil and Gas Investor. http://www.hartenergy.com.

He, B. 2004. The Effect of Metallic Salt Additives on In-Situ Combustion Performance. M.S. Report. Stanford University, Stanford, California. http://ekofisk.stanford.edu/pereports/web/default.htm

He, B., Chen, Q., Castanier, L.M., and Kovscek, A.R. 2005. Improved In-Situ Combustion Performance with Metallic Salt Additives. Paper No. SPE 93901. Proceedings of the Society of Petroleum Engineers Western Regional Meeting. Irvine, California, March 30–April 1.

Hertzberg, R., Hojabri, F., and Ellefson, L. 1983. Preprint No. 35e. Summer National Meeting of the American Institute of Chemical Engineers. Denver, Colorado, August 28–31.

Howard, F.A. 1923. Method of Operating Oil Wells. U.S. Patent No. 1,473,348.

IEA. 2005. Resources to Reserves—Oil and Gas Technologies for the Energy Markets of the Future. International Energy Agency, Paris. www.iea.org/Textbase/publications/free_new_Desc.asp?PUBS_ID=1568.

Jiang, S., Liu, X., Liu, Y., and Zhong, L. 2005. In Situ Upgrading Heavy Oil by Aquathermolytic Treatment under Steam Injection Conditions. SPE.

Johnson, L.A., Fahy, L.J., Romanowski, L.J., Barbour, R.V., and Thomas, K.P. 1980. An Echoing In-Situ Combustion Oil Recovery Project in a Utah Tar Sand. Journal of Petroleum Technology. February:295–305.

Koch, R.L. 1956. Practical Use of Combustion Drive at West Newport Field. Petroleum Engineering. January:72.

Kuhn, C.S., and Koch, R.L. 1953. In-Situ Combustion—Newest Method of Increasing Oil Recovery. Oil and Gas Journal. 52(14):92.

Lake, L.W., Schmidt, R.L., and Venuto, P.B. 1992. A Niche for Enhanced Oil Recovery in the 1990s. Oilfield Review, January:55–61.

Lake, L.W., and Walsh, M.P. 2004. Primary Hydrocarbon Recovery. Elsevier, Amsterdam, The Netherlands.

Martin, W.L., Alexander, J.D., and Dew, J.N. 1958. Process Variables of In-Situ Combustion. Trans., AIME. (1958) 213, 28–35.

Meszaros, G., Chakma, A., Zha, K. N., and Islam, M. R. 1990. Scaled Model Studies and Numerical Simulation of Inert Gas Injection with Horizontal Wells. Paper No. SPE 20529. Proceedings. 65th Society of Petroleum Engineers Annual Technical Conference and Exhibition. New Orleans, Louisiana, September 23–26.

Mitchell, D.L., and Speight, J.G. 1973. The Solubility of Asphaltenes in Hydrocarbon Solvents. Fuel. 52:149.

Mut, S. 2005. Advances in an In Situ Upgrading Process for Unconventional Oils. American Association of Petroleum Geologists International Conference and Exhibition. Paris, France, September 11–14.

Northrop, P.S., and Venkatesan, V.N. 1993. Analytical Steam Distillation Model for Thermal Enhanced Oil Recovery Processes. Industrial and Engineering Chemistry. Research. 32(9):2039–2046.

Parrish, D.R., and Craig, F.F. Jr. 1969. Laboratory Study of a Combination of Forward Combustion and Waterflooding—The COFCAW Process. Journal of Petroleum Technology. 21:753–761.

Prats, M. 1986. Thermal Recovery. Society of Petroleum Engineers, Richardson, Texas.

Ramey, H.J. Jr. 1971. In-Situ Combustion. Proceedings, 253–262. World Petroleum Congress, Moscow.

Ramey, H.J., Jr, Stamp, V.V., and Pebdani, F.N. 1992. Case History of South Belridge, California, In-Situ Combustion Oil Recovery. Paper No. SPE 24200. Proceedings. 9th Society of Petroleum Engineers/US Department of Energy Enhanced Oil Recovery Symposium. Tulsa, Oklahoma. April 21–24.

Reichert, C., Fuhr, B., Sawatzky, H., Lafleur, R., Verkoczy. B., Soveran, D., and Jha, K. 1989. Petroleum Science and Technology. 7:851–878.

Richardson, W.C., Fontaine, M.F., and Haynes, S. 1992. Paper No. SPE 24033. Western Regional Meeting. Bakersfield, California, March 30–April 1.

Shallcross, D.C., Rios, C.F. De Los, Castanier, L.M. 1991. Modifying In-Situ Combustion Performance by the Use of Water-Soluble Additives. SPERE (Aug.):287–294.

Speight, J.G. 2007. The Chemistry and Technology of Petroleum. 4th Edition. CRC Press, Taylor and Francis Group, Boca Raton, Florida.

Strycker, A., Sarathi, P., and Wang, S., 1999. Evaluation of In Situ Combustion for Schrader Bluff. Topical Report, National Petroleum

Technology Office, United States Department of Energy, Washington, DC. http://www.osti.gov/bridge/.

Tadema, H.J., and Weijdema, J. 1970. Spontaneous Ignition in Oil Sands. Oil and Gas Journal. December:77–80.

Watts, K.C., Hutchinson, H.L., Johnson, L.A., Barbour, R.V., and Thomas, K.P. 1982. Proceedings. 54th Annual Fall Meeting, Society of Petroleum Engineers, American Institute of Mechanical Engineers. New Orleans. September 26–29.

Wilson, L.A., and Root, P.J. 1966. Cost Comparison of Reservoir Heating Using Steam or Air. Journal of Petroleum Technology. 18:233–239.

Winestock, A.G. 1974. Oil Sands Fuel of the Future, ed. L. V. Hills, 190. Canadian Society of Petroleum Geologists, Calgary, Alberta, Canada.

Wolcott, E.R. 1923. Method of Increasing the Yield of Oil Wells. U.S. Patent No. 1,457,479.

Wu, C.H. and Fulton, P.F. 1971. Experimental Simulation of the Zones Preceding the Combustion Front of an In-Situ Combustion Process. Society of Petroleum Engineers Journal. March:38–46.

Xia, T.X., Greaves, M., Werfilli, W.S., and Rathbone, R.R. 2002. Downhole Conversion of Lloydminster Heavy Oil Using THAI-CAPRI Processes. SPE 78998.

Yang, C., and Gu, Y. 2005a A Novel Experimental Technique for Studying Solvent Mass Transfer and Oil Swelling Effect in a Vapor Extraction (VAPEX) Process. Paper No. 2005-099. Proceedings. 56th Annual Technical Meeting. The Canadian International Petroleum Conference. Calgary, Alberta, Canada, June 7–9.

Yang, C., and Gu, Y. 2005b. Effects of Solvent-Heavy Oil Interfacial Tension on Gravity Drainage in the VAPEX Process. Paper No. SPE 97906. Society of Petroleum Engineers International Thermal Operations and Heavy Oil Symposium. Calgary, Alberta, Canada. November 1–3.

CHAPTER 8

UPGRADING HEAVY OIL

Upgrading heavy crude oil is of major economic importance (Hedrick et al., 2006). Heavy crude oils exist in large quantities in the Western Hemisphere, but they are difficult to produce and transport because of their high viscosity. Some crude oils contain compounds such as sulfur and/or heavy metals, which cause additional refining problems and costs. In situ upgrading could be a very beneficial process for leaving the unwanted elements in the reservoir and increasing API gravity.

Fluids produced from a well are seldom pure crude oil. In fact, a variety of materials may be produced by oil wells in addition to liquid and gaseous hydrocarbons. The natural gas itself may contain as impurities one or more non-hydrocarbon substances. The most abundant of these impurities is hydrogen sulfide, which imparts a noticeable odor to the gas. A small amount of this compound is considered advantageous as it gives an indication of leaks and where they occur. A larger amount, however, makes the gas obnoxious and difficult to market. Such gas is referred to as *sour gas* (Chapter 1), and much of it is used in the manufacture of carbon black. A few natural gases contain helium, and this element does in fact occur in commercial quantities in certain gas fields. Nitrogen and carbon dioxide are also found in some natural gases. Gas is usually separated at the highest pressure possible, which reduces compression costs when the gas is to be used for gas lift or delivered to a pipeline. Lighter hydrocarbons and hydrogen sulfide are removed as necessary to obtain a crude oil of suitable vapor pressure for transport that retains most of the natural gasoline constituents.

By far, the most abundant extraneous material is water. Many wells, especially during their declining years, produce vast quantities of salt water, and disposing of it is both a serious and expensive problem. Furthermore, the brine may be corrosive, which necessitates frequent replacement of casing, pipe, and valves, or it may be saturated so that the salts tend to precipitate upon reaching the surface. In either case, the water produced with the oil is a source of continuing trouble.

Finally, if the reservoir rock is an incoherent sand or poorly cemented sandstone, large quantities of sand are produced along with the oil and gas. On its way to the surface, the sand has been known to scour its way completely through pipes and fittings.

It must also be remembered that in any field where primary production is followed by a secondary or enhanced production method, there will be noticeable differences in properties between the fluids produced (Thomas et al., 1983). These differences may not be reflected in the elemental composition to any great extent (Zou et al., 1989), but they will be evident from an inspection of the physical properties. One issue that arises from the physical property data is that such oils may be outside the range of acceptability for refining techniques other than thermal options. In addition, overloading of thermal process units will increase as the proportion of the heavy oil in the refinery feedstock increases. Obviously, there is a need for more and more refineries to accept larger proportions of heavy crude oils as refinery feedstock and to have the capability to process such materials.

Technologies such as alkaline flooding, microemulsion (micellar/emulsion) flooding, polymer augmented waterflooding, and carbon dioxide miscible/immiscible flooding do not require or cause any change to the oil. The steaming technologies may cause some steam distillation that can augment oil recovery when the steam distilled material moves with the steam front and acts as a solvent for oil ahead of the steam front (Pratts, 1986). Again, there is no chemical change to the oil although there may be favorable compositional changes to the oil insofar as lighter fractions are recovered and heavier materials remain in the reservoir (Richardson et al., 1992).

The mature and well-established processes such as visbreaking, delayed coking, fluid coking, flexicoking, propane deasphalting, and butane deasphalting were deemed adequate for upgrading heavy feedstocks. More options are now being sought in order to increase process efficiency in terms of the yields of the desired products. It is the purpose of this chapter to present an outline of the options for (1) surface upgrading and (2) for in situ upgrading.

8.1 Surface Upgrading

The influx of heavy oils into the refinery system can offset the shortages of conventional crude oil, but there is also a need for increased refining capacity. New residue-processing capacity needs be added to existing refineries or be built in separate, stand-alone upgrading facilities. If the oil is too viscous to transport by pipeline and/or there is the need for heat or energy at the production site, heavy oil upgrading in the field is attractive. It may also avoid extensive modifications of existing refineries.

Petroleum refining is now in a significant transition period as the industry moves further into the 21st century. The demand for petroleum and petroleum products has shown a sharp growth in recent decades. In order to satisfy the changing pattern of product demand, significant investments in refining conversion processes will be necessary to profitably utilize heavy feedstocks. The most efficient and economical solutions to this problem will depend to a large extent on individual refinery situations. However, the most promising technologies will likely involve the conversion of vacuum residua. Technologies are also needed that will take the feedstock beyond current limits and, at the same time, reduce the amount of coke and other nonessential products. Such a goal may require the use of two or more technologies in series rather than a whole new one-stop conversion technology. Such is the nature of the refinery.

A refinery is an integrated collection of unit processes (Figure 8–1); it can have any one or a conjunction of several configurations. The refinery of the future, however, will be required to be a *conversion refinery*. A conversion refinery incorporates all the basic building blocks found in both the *topping refinery* and the *hydroskimming refinery*, and it also features gas oil conversion plants such as catalytic cracking and hydrocracking units, olefin conversion plants such as alkylation or polymerization units, and (frequently) coking units for sharply reducing or eliminating the production of residual fuels. Conversion refineries currently produce as much as two thirds of their output as unleaded gasoline, with the balance distributed between liquefied petroleum gas (LPG), high-quality jet fuel, low-sulfur diesel fuel, and coke. Many such refineries also incorporate solvent extraction

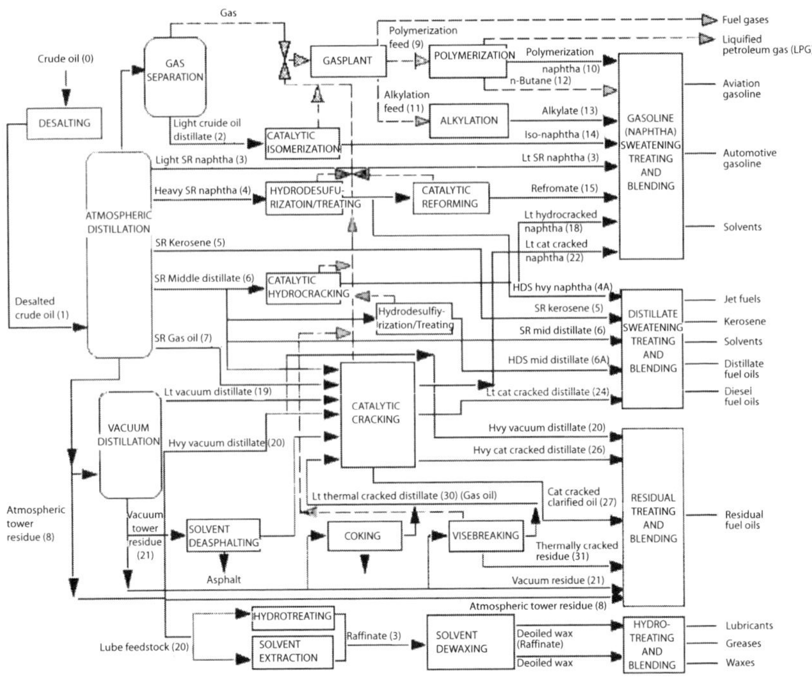

Figure 8–1 *A conventional refinery.*

processes for manufacturing lubricants and petrochemical units with which to recover high-purity propylene, benzene, toluene, and xylenes for further processing into polymers.

The manner in which refineries convert heavy oil into low-boiling high-value products has become a major focus of operations, with new concepts evolving into new processes (Khan and Patmore, 1998; Speight, 2000, 2007). Even though they may not be classed as conversion processes per se, pretreatment processes for removing asphaltene constituents, metals, sulfur, and nitrogen constituents are also important and can play an important role.

New processes for the conversion of residua and heavy oils will probably be used not only in place of but also in conjunction with visbreaking and coking options with some degree of hydroprocessing as a primary conversion step. In addition, other processes may replace or (more likely) augment the deasphalting units in many refineries (Speight, 1990, 2000, 2007).

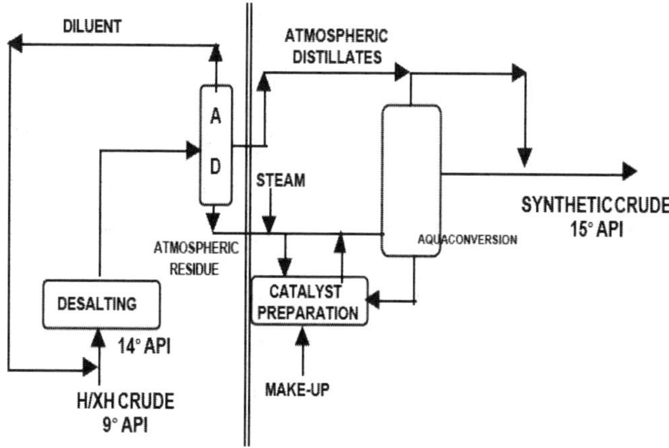

Figure 8–2 *The Aquaconversion process.*

Aquaconversion technology (Figure 8–2) offers a novel way of upgrading heavy crude oil. It is a catalytic hydrovisbreaking process that operates in the presence of steam. Visbreaking technology is limited in conversion level because of the stability of the resulting product. Because one process requirement is that the synthetic crude oil (syncrude) has to be stable, standard visbreaking allows only a 2° API upgrading of the heavy crude oil and only a limited viscosity reduction, which does not ensure its transport without external diluent. The aquaconversion process pushes this maximum conversion level within the stability specification by adding a homogeneous catalyst in the presence of steam (Marzin et al., 1998).

This novel catalytic system allows hydrogen from the water to be transferred to the resid when operated at the conditions normally used for the visbreaking process. Similar operating conditions (pressure and temperature) are used. Hydrogen incorporation is much lower than that obtained when using a deep hydroconversion process under high hydrogen partial pressure. Nevertheless, it is high enough to saturate the reactive species formed within the thermal process that would normally lead to the formation of coke precursors and coke. Because of hydrogen incorporation, a higher conversion level can be reached, which enables higher API and viscosity improvements to be achieved while maintaining product stability.

Important aspects of Aquaconversion technology is that it does not produce any solid by-product such as coke nor require any hydrogen source or high-pressure equipment. In addition, the Aquaconversion process can be implanted in the production area, and thus the need for external diluent and its transport over large distances is eliminated. Light distillates from the raw crude oil can be used as diluent for both the production and desalting processes.

Conceivably, heavy oil could be upgraded at the wellhead by a primary upgrading (in fact, a partial upgrading) step, such as Aquaconversion. Upgrading heavy oil in the future will adapt to the nature of the recovered oil by an as-yet-to-be-specified upgrading step at the recovery stage. Technologies that are adequate for this task will need to be identified, and the level of upgrading will also need to be determined. The use of visbreaking, hydrovisbreaking, or deasphalting are options for such an upgrading step. Indeed, partial upgrading as part of the recovery process may be a determinant of the price differentials (Fattouh, 2006) between the recovered oil and a benchmark crude oil.

Visbreaking and its variants and the recently demonstrated ORMAT process, which is a combination of visbreaking and deasphalting, offer milder thermal and physical processes that can play one of two possible roles: (1) as field upgraders to reduce or eliminate the need for diluent for transport or (2) as primary processes in future upgraders where some residue is removed and consumed in hydrogen production (Gunter et al., 2005).

On the other hand, there are a number of technologies and technology concepts that have been described since the 1980s that are believed to be suitable for upgrading in the refinery (Table 8–1). These technologies are in addition to the more standard and recognizable processes such as coking and hydrocracking technologies (Moschopedis et al., 1998; Speight and Ozum, 2002; Speight, 2007). Coking technologies can be carbon wasteful in terms of the amount of coke produced (when the goal is to produce liquid fuels), and hydrocracking technologies do not fare much better since heavy oil feedstocks can have a serious affect on catalyst activity and life. In addition, blending heavy oil with lighter oil may not always be an option for introduction of heavy oil to the refinery. Shortages of lighter oil(s) or incompatibility of the heavy oil with the light oil may be limitations to the blending option.

However, it is not abundantly clear if the newer process concepts will change not only the quantity of the product but also, sometimes inadvertently, change the quality of the product. While an upward trend in product yield is usually looked upon as a positive change, a change in product quality is not always a positive change, and it may lead to a reevaluation of the steps needed to produce a specification product for sales.

Nevertheless, the alternate processes are what are described in the following sections. They could be applicable to heavy oil upgrading as described or, depending upon feedstock properties, in a modified form.

8.1.1 Thermal Cracking Processes

Thermal cracking processes offer attractive methods of feedstock conversion at low operating pressure without requiring expensive catalysts. Currently, the widest-operated residuum conversion processes are visbreaking, delayed coking, and fluid coking, which are still attractive processes for refineries from an economic point of view (Dickenson et al., 1997).

The *ASCOT process* is a residual oil upgrading process that integrates the delayed coking process and the deep solvent deasphalting process (low energy deasphalting, LEDA) (Bonilla, 1985; Bonilla and Elliot, 1987; RAROP, 1991, p. 3; Hydrocarbon Processing, 1996). The product yields can be varied according to the desired distribution of products by emphasizing one part or the other or both parts of the process.

In the process, the feedstock is brought to the desired extraction temperature (50°C to 230°C, 120°F to 445°F, at 300 to 500 psig, 2060 to 3430 kPa) and then sent to the extractor, where solvent (straight run naphtha, coker naphtha; solvent to oil ratio = 4:1 to 13:1) flows upward, extracting soluble material from the down-flowing feedstock. The solvent-deasphalted phase leaves the top of the extractor and flows to the solvent recovery system, where the solvent is separated from the deasphalted oil and recycled to the extractor. The deasphalted oil is sent to the delayed coker (heater outlet temperature: 480°C to 510°C, 900°F to 950°F, at 15 to 35 psig, 105 kPa to 240 kPa, and a recycle ratio of 0.05 to 0.25 on fresh feedstock), where it is combined with the heavy coker gas oil from the coker fractionator. This mixture is sent to the heavy coker gas oil stripper, where low-boiling hydrocarbons are stripped off and returned to the fractionator.

The stripped deasphalted oil/heavy coker gas oil mixture is removed from the bottom of the stripper and used to provide heat to the naphtha stabilizer-reboiler before being sent to battery limits as a cracking stock. The raffinate phase containing the asphalt and some solvent flows at a controlled rate from the bottom of the extractor and is charged directly to the coking section.

The *comprehensive heavy ends reforming refinery (Cherry-P) process* is a process for the conversion of heavy crude oil or residuum into distillate and a cracked residuum (Ueda, 1976, 1978; RAROP, 1991, p. 5). In the process, the feedstock is mixed with coal powder in a slurry-mixing vessel, heated in the furnace, and fed to the reactor, where the feedstock undergoes thermal cracking reactions for several (three to five) hours at a temperature higher than 400°C to 430°C (750°F to 805°F) and under pressure of 140 to 280 psig (980 to 1960 kPa). Gas and distillate from the reactor are sent to a fractionator, and the cracked residuum is extracted out of the system after distilling low-boiling fractions by the flash drum and vacuum flasher to adjust its softening point. The distillates produced by this process are generally lower in the content of olefin hydrocarbons than the other thermal cracking processes, comparatively easy to desulfurize in hydrotreating units, and compatible with straight-run distillates.

The *deep thermal conversion (DTC) process* offers a bridge between visbreaking and coking and provides maximum distillate yields by applying deep thermal conversion to vacuum residua followed by vacuum flashing of the products (Hydrocarbon Processing, 1998, p. 69). In the process, the heated feedstock is charged to the heater and from there to the soaker where conversion occurs. The products are then led to an atmospheric fractionator to produce gases, naphtha, kerosene, and gas oil. The fractionator residuum is sent to a vacuum flasher that recovers additional gas oil and distillate. The next steps for the coke are dependent on its potential use; it may be isolated as *liquid* coke (pitch, cracked residuum) or solid coke.

The *ET-II process* is a thermal cracking process for the production of distillates and cracked residuum for use as a metallurgical coke and is designed to accommodate feedstocks such as heavy oils, atmospheric residua, and vacuum residua (Kuwahara, 1987; RAROP, 1991, p. 9). The distillate (referred to in the process as *cracked oil*) is suitable as a feedstock to hydrocracking and fluid catalytic cracking. The basic technology of the ET-II process is derived from that of the original Eureka process. In the ET-II process, the feedstock is heated up to 350°C (660°F) by passage through the preheater and fed into the bottom of the fractionator, where it is mixed with recycle oil, the high-boiling fraction of the cracked oil. The ratio of recycle oil to feedstock is within the range (wt. %) of 0.1 to 0.3. The feedstock mixed with recycle oil is then pumped out and fed into the cracking heater, where the temperature is raised to approximately 490°C to 495°C (915-925°F), and the outflow is fed to the stirred-tank reactor where it is subjected to further thermal cracking. Both cracking and condensation reactions take place in the reactor.

The heat required for the cracking reaction is brought in by the effluent itself from the cracking heater as well as by the superheated steam, which is heated in the convection section of the cracking heater and blown into the reactor bottom. The superheated steam reduces the partial pressure of the hydrocarbons in the reactor and accelerates the stripping of volatile components from the cracked residuum. This residual product is discharged through a transfer pump and transferred to a cooling drum, where the thermal cracking reaction is terminated by quenching with a water spray. After this, it is sent to the pitch water slurry preparation unit. The cracked oil and gas products, together with steam from the top of the reactor, are introduced into the fractionator, where the oil is separated into three fractions, *cracked light oil*, *cracked heavy oil*, and *cracked residuum* (*pitch*).

The *Eureka process* is a thermal cracking process to produce a cracked oil and aromatic residuum from heavy residual materials (Aiba et al., 1981; RAROP, 1991, p. 11; Chen, 1993). In this process, the feedstock (usually a feedstock) is fed to the preheater and then enters the bottom of the fractionator, where it is mixed with the recycle oil. The mixture is then fed to the reactor system that consists of a pair of reactors operating alternately. In the reactor, a thermal cracking reaction occurs in the presence of superheated steam, which is injected to strip the cracked products out of the reactor and to supply a part of heat required for cracking reaction. At the end of the reaction, the bottom product is quenched.

The oil and gas products (and steam) pass from the top of the reactor to the lower section of the fractionator, where a small amount of entrained material is removed by a wash operation. The upper section is an ordinary fractionator, where the heavier fraction of cracked oil is drawn as a side stream.

The *fluid thermal cracking (FTC) process* is a heavy oil and residuum upgrading process in which the feedstock is thermally cracked to produce distillate and coke, which is gasified to fuel gas (Miyauchi et al.,

1981; Miyauchi and Ikeda, 1988; RAROP, 1991, p. 17). The feedstock, mixed with recycle stock from the fractionator, is injected into the cracker, is immediately absorbed into the pores of the particles by capillary force, and is subjected to thermal cracking. In consequence, the surface of the noncatalytic particles is kept dry, and good fluidity is maintained, allowing a good yield of, and selectivity for, middle distillate products. Hydrogen-containing gas from the fractionator is used for the fluidization in the cracker.

Excessive coke caused by the metals accumulated on the particles is suppressed under the presence of hydrogen. The particles with deposited coke from the cracker are sent to the gasifier, where the coke is gasified and converted into carbon monoxide (CO), hydrogen (H_2), carbon dioxide (CO_2), and hydrogen sulfide (H_2S) with steam and air. Regenerated hot particles are returned to the cracker.

The *high conversion soaker cracking (HSC) process* is a cracking process to cover a moderate conversion, higher than visbreaking but lower than coking to various products (Watari et al., 1987; Washimi, 1989; RAROP, 1991, p. 19; Washimi and Hamamura, 1993). The preheated feedstock enters the bottom of the fractionator, where it is mixed with the recycle oil. The mixture is pumped up to the charge heater and fed to the soaking drum (ca. atmospheric pressure, steam injection at the top and bottom), where sufficient residence time is provided to complete the thermal cracking. In the soaking drum, the feedstock and some product flows downward, passing through a number of perforated plates while steam with cracked gas and distillate vapors flow through the perforated plates countercurrently.

The volatile products from the soaking drum enter the fractionator, where the distillates are fractionated into desired product oil streams, including a heavy gas oil fraction. The cracked gas product is compressed and used as refinery fuel gas after sweetening. The cracked oil product after hydrotreating is used as fluid catalytic cracking or hydrocracking feedstock. The residuum is suitable for use as boiler fuel, road asphalt, binder for the coking industry, and as a feedstock for partial oxidation.

In the *Tervahl T process* (LePage et al., 1987; RAROP, 1991, p. 25), the feedstock is heated to the desired temperature using the coil heater and heat recovered in the stabilization section and is held for a specified residence time in the soaking drum. The soaking drum effluent is quenched and sent to a conventional stabilizer or fractionator, where the products are separated into the desired streams. The gas produced from the process is used for fuel.

In the *Tervahl H process* (Speight, 2007) the feedstock and hydrogen-rich stream are heated using heat recovery techniques and a fired heater and held in the soak drum as in the Tervahl T process. The gas and oil from the soaking drum effluent are mixed with recycle hydrogen and separated in the hot separator, where the gas is cooled, passed through a separator, and recycled to the heater and soaking drum effluent. The liquids from the hot and cold separator are sent to the stabilizer section, where purge gas and synthetic crude are separated. The gas is used as fuel, and the synthetic crude can now be transported or stored.

8.1.2 Catalytic Cracking Processes

The *fluid catalytic cracking process*, using vacuum gas oil as the feedstock, was introduced into refineries in the 1930s. In recent years, because of a trend for low-boiling products, most refineries have performed the operation by partially blending residua into vacuum gas oil. However, conventional fluid catalytic cracking processes have limits in residuum processing, so residuum fluid catalytic cracking processes have lately been employed one after another. Because the residuum fluid catalytic cracking process enables efficient gasoline production directly from residua, it will play the most important role as a residuum cracking process, separately or in conjunction with a residuum hydrotreating process (Reynolds et al., 1992).

The processes described in the following paragraphs are the evolutionary offspring of the fluid catalytic cracking and the residuum catalytic cracking processes. Some of these newer processes use catalysts with different silica/alumina ratios as acid support of metals such as molybdenum (Mo), cobalt (Co), nickel (Ni), and tungsten (W). In general, the first catalyst used to remove metals from oils was the conventional hydrodesulfurization (HDS) catalyst. Diverse natural minerals are also used as raw material for elaborating catalysts addressed to the upgrading of heavy fractions. Among these minerals are clays; manganese nodules; bauxite activated with vanadium (V), nickel (Ni), chromium (Cr), iron (Fe), and cobalt (Co), as well as iron laterites and sepiolites; and mineral nickel and transition metal sulfides supported on silica and alumina. Other kinds of catalysts, such as vanadium sulfide, are generated in situ, possibly in colloidal states.

The *asphalt residua treating (ART) process* is a process for increasing the production of high value distillates (Logwinuk and Caldwell, 1983; Green and Center, 1985; Gussow and Kramer, 1990; RAROP, 1991, p. 29; Bartholic et al., 1992; Hydrocarbon Processing, 1996). In the process, the preheated feedstock (which may be whole crude oil, atmospheric residuum, feedstock, or bitumen) is injected into a stream of fluidized, hot catalyst (trade name: ArtCat). The unit configuration is similar to that of a riser fluid catalytic cracking unit, where complete mixing of the feedstock with the catalyst is achieved in the contactor, which is operated within a pressure-temperature envelope to ensure selective vaporization. The vapor and the contactor effluent are quickly and efficiently separated from each other, and entrained hydrocarbons are stripped from the contaminant (containing spent solid) in the stripping section. The contactor vapor effluent and vapor from the stripping section are combined and rapidly quenched in a quench drum to minimize product degradation. The cooled products are then transported to a conventional fractionator that is similar to that found in a fluid catalytic cracking unit. Spent solid from the stripping section is transported to the combustor bottom zone for carbon burn-off.

Contact of the feedstock with the fluidizable catalyst in a short-residence-time contactor causes the lower-boiling components of the feedstock to vaporize, and asphaltene constituents (high-molecular-weight compounds) to crack to yield lower-boiling compounds and coke. The metals present, as well as some of the sulfur and the nitrogen compounds in the nonvolatile constituents, are retained on the catalyst. At the exit of the contacting zone, the oil vapors are separated from the catalyst and are rapidly quenched to minimize thermal cracking of the products. The catalyst, which holds metals, sulfur, nitrogen, and coke, is transferred to the regenerator, where the combustible portion is oxidized and removed. Regenerated contact material, bearing metals but very little coke, exits the regenerator and passes to the contactor for further removal of contaminants from the charge stock.

In the combustor, coke is burned from the spent solid, which is then separated from combustion gas in the surge vessel. The surge vessel circulates regenerated catalyst streams to the contactor inlet for feed vaporization and to the combustor bottom zone for premixing. The components of the combustion gases include carbon dioxide (CO_2), nitrogen (N_2), oxygen (O_2), sulfur oxides (SO_X), and nitrogen oxides (NO_X), which are released from the catalyst with the combustion of the coke in the combustor. The concentration of sulfur oxides in the combustion gas requires treatment for their removal.

The *heavy oil treating (HOT) process* is a catalytic cracking process for upgrading heavy feedstocks such as topped crude oils, vacuum residua, and solvent-deasphalted bottoms using a fluidized bed of iron ore particles (Ozaki, 1982; RAROP, 191, p. 35). The main section of the process consists of three fluidized reactors. Separate reactions take place in each reactor (*cracker, regenerator,* and *desulfurizer*):

Fe_3O_4 + asphaltenes → coke/Fe_3O_4 + Oil + Gas (in the *cracker*)

3FeO + H_20 → Fe_3O_4 + H_2 (in the *cracker*)

coke/Fe_3O_4 + O_2 → 3FeO + CO + CO_2 (in the *regenerator*)

FeO + SO_2 + 3CO → FeS + 3CO_2 (in the *regenerator*)

3FeS + 5O_2 → Fe_3O_4 + 3$S0_2$ (in the *desulfurizer*)

In the *cracker,* heavy oil cracking and the steam-iron reaction take place simultaneously under the conditions usual to thermal cracking. Any unconverted feedstock is recycled to the cracker from the bottom of the scrubber. The scrubber effluent is separated into hydrogen gas, liquefied petroleum gas (LPG), and liquid products that can be upgraded by conventional technologies to priority products. In the *regenerator,* coke deposited on the catalyst is partially burned to form carbon monoxide, which reduces iron tetroxide and acts as a heat supply. In the *desulfurizer,* sulfur in the solid catalyst is removed and recovered as molten sulfur in the final recovery stage.

The *R2R process* is a fluid catalytic cracking process for conversion of heavy feedstocks (RAROP, 1991, p. 37). In the process, the feedstock is vaporized upon contacting hot regenerated catalyst at the base of the riser. The vaporized feedstock lifts the catalyst into the reactor vessel separation chamber, where rapid disengagement of the hydrocarbon vapors from the catalyst is accomplished by a special solids separator and cyclones. The bulk of the cracking reaction takes place at the moment of contact and continues as the catalyst and hydrocarbons travel up the riser. The reaction products, along with a minute amount of entrained catalyst, then flow to the fractionation column. The stripped spent catalyst, deactivated with coke, flows into the Number 1 regenerator. Partially regenerated catalyst is pneumatically transferred via an air riser to the Number 2 regenerator, where the remaining carbon is completely burned in a dryer atmosphere.

In the *reduced crude oil conversion (RCC)* process, the clean regenerated catalyst enters the bottom of the reactor riser where it contacts low-boiling hydrocarbon *lift gas* that accelerates the catalyst up the riser prior to feed injection (Hydrocarbon Processing, 1996; RAROP, 1991, p. 39). At the top of the lift gas zone, the feed is injected through a series of nozzles located around the circumference of the reactor riser. The catalyst/oil disengaging system is designed to separate the catalyst from the reaction products and then rapidly remove the reaction products from the reactor vessel. Spent catalyst from the reaction zone is first steam stripped to remove adsorbed hydrocarbon and then routed to the regenerator. In the regenerator, all of the carbonaceous deposits are removed from the catalyst by combustion, restoring the catalyst to an active state with a very low carbon content. The catalyst is then returned to the bottom of the reactor riser at a controlled rate to achieve the desired conversion and selectivity to the primary products.

The *residue fluid catalytic cracking (HOC) process* is a version of the fluid catalytic cracking process that has been adapted to conversion of a wide range of feedstocks that contain high amounts of metal and asphaltenes (Finneran, 1974; Murphy and Treese, 1979; Johnson, 1982; RAROP, 1991, p. 33; Feldman et al., 1992). In the process, the flow of solids is essentially vertical. Regenerator-bed temperatures are limited to around 730°C (1,300°F), and feed-introduction systems are designed for efficient mixing of oil and catalyst and rapid quenching of the catalyst temperature to the equilibrium mix temperature. The reaction system is an external vertical riser providing very low contact times and terminating in the riser cyclones for rapid separation of catalyst and vapors. A two-stage stripper is utilized to remove hydrocarbons from the catalyst. Hot catalyst flows at low velocity in dense phase through the catalyst cooler and returns to the regenerator. Regenerated catalyst flows to the bottom of the riser to meet the feed.

The coke deposited on the catalyst is burned off in the regenerator along with the coke formed during the cracking of the gas oil fraction. The high amounts of coke produced in cracking of residua can cause extreme temperatures and excessive catalyst deactivation. Steam coils located within the regenerator bed and/or external catalyst coolers remove the excess heat produced by the high coke yields. Depending on quality and product objectives, feedstocks with vanadium-plus-nickel content of up to 30 ppm and carbon residue of up 10% by weight can be processed without feed pretreatment. When the metals content and carbon residua data are in excess of those values, the feedstock may require pretreatment. If the feedstock contains high proportions of metals, control of the metals on the catalyst requires excessive amounts of catalyst withdrawal and fresh catalyst addition. This problem can be addressed by feedstock pretreatment, such as hydrodesulfurization.

Hydrovisbreaking, a non-catalytic process, is conducted under similar conditions to visbreaking and involves treatment with hydrogen under mild conditions (RAROP, 1991, p. 57). The presence of hydrogen leads to more stable products (lower flocculation threshold) than can be obtained with straight visbreaking. This means that higher conversions can be achieved, which produce a lower viscosity product.

The *HYCAR process* is composed fundamentally of three parts: (1) visbreaking, (2) hydrodemetallization, and (3) hydrocracking. In the visbreaking section, the heavy feedstock is subjected to moderate thermal cracking while no coke formation is induced. The visbroken oil is fed to the demetallization reactor in the presence of catalysts, which provide sufficient pore for diffusion and adsorption of high-molecular-weight constituents. The product from this second stage proceeds to the hydrocracking reactor, where desulfurization and denitrogenation take place along with hydrocracking.

8.1.3 Hydrogen Addition Processes

Hydrotreating processes for heavy oils have three definite roles: (1) desulfurization to supply low-sulfur fuel oils, (2) pretreatment of the feedstock for fluid catalytic cracking processes, and (3) hydrocracking to produce feedstocks for fluid catalytic cracking processes. The hydrotreating processes are becoming more popular as pretreating processes where the main goal is to remove sulfur, metal, and asphaltene contents from residua and other heavy feedstocks to a desired level and, at the same time, maintain hydrogen consumption at acceptable levels that are dictated, for example, by the degree of desulfurization and/or hydrocracking (Maples, 2000; Speight, 2000). Hydrocracking process options for residua are also seeing an increase in use but have been considered in the past *hydrogen sinks* (being somewhat uncontrollable and therefore wasteful in hydrogen use). They are, therefore, seemingly less popular than the hydrotreating processes. Hopefully this attitude is changing as more hydroconversion processes are being introduced for heavy feed conversion.

The major goal of *residuum hydroconversion* is the cracking of residua with desulfurization, metal removal, denitrogenation, and asphaltene conversion. The residuum hydroconversion process offers production of kerosene and gas oil and production of feedstocks for hydrocracking, fluid catalytic cracking, and petrochemical applications.

The *asphaltenic bottom cracking (ABC) process* can be used for hydrodemetallization, asphaltene cracking, and moderate hydrodesulfurization, as well as sufficient resistance to coke fouling and metal deposition using such feedstocks as vacuum residua, thermally cracked residua, solvent deasphalted bottoms, and bitumen with fixed catalyst beds (Takeuchi, 1982; RAROP, 1991, p. 45). In the process, the feedstock is pumped up to the reaction pressure and mixed with hydrogen. The mixture is heated to the reaction temperature in the charge heater after a heat exchange and fed to the reactor. In the reactor, hydrodemetallization and subsequent asphaltene cracking with moderate hydrodesulfurization take place simultaneously under conditions similar to residuum hydrodesulfurization. The reactor effluent gas is cooled, cleaned up, and recycled to the reactor section, while the separated liquid is distilled into distillate fractions and feedstock, which is further separated into deasphalted oil and asphalt using butane or pentane. In case of the ABC-hydrodesulfurization catalyst combination, the ABC catalyst is placed upstream of the hydrodesulfurization catalyst and can be operated at a higher temperature than the hydrodesulfurization catalyst under conventional residuum hydrodesulfurization conditions. In the VisABC process, a soaking drum is provided after the heater, when necessary. Hydrovisbroken oil is first stabilized by the ABC catalyst through hydrogenation of coke precursors and then desulfurized by the HDS catalyst.

The *Hyvahl F process* is used to hydrotreat atmospheric and vacuum residua to convert the feedstock to naphtha and middle distillates (Hydrocarbon Processing, 1996; RAROP, 1991, p. 59). The main features of this process are its dual catalyst system and its fixed-bed swing-reactor concept. The first catalyst has a high capacity for metals (to 100% w/w of new catalyst) and is used for both hydrodemetallization (HDM) and most of the conversion. This catalyst is resistant to fouling, coking, and plugging by asphaltenes and shields the second catalyst from the same. Protected from metal poisons and deposition of cokelike products, the highly active second catalyst can carry out its deep HDS and refining functions. Both catalyst systems use fixed beds that are more efficient than moving beds and are not subject to attrition problems.

In the process, the preheated feedstock enters one of the two guard reactors, where a large proportion of the nickel and vanadium are adsorbed, and hydroconversion of the heavy molecule weight constituents commences. Meanwhile, the second guard reactor catalyst undergoes a reconditioning process and then is put on standby. From the guard reactors, the feedstock flows through a series of hydrodemetallization reactors that continue the metals removal and the conversion of heavy ends. The next processing stage, hydrodesulfurization, is where most of sulfur, some of the nitrogen, and the residual metals are removed. A limited amount of conversion also takes place. From the final reactor, the gas phase is separated, hydrogen is recirculated to the reaction section, and the liquid products are sent to a conventional fractionation section for separation into naphtha, middle distillates, and heavier streams.

8.1.4 Solvent Processes

Solvent deasphalting processes have not realized their maximum potential in terms of use with heavy oil. With ongoing improvements in energy efficiency, such processes would display their effects in combination with other processes. Solvent deasphalting allows removal of sulfur and nitrogen compounds as well as metallic constituents by balancing yield with the desired feedstock properties (Ditman, 1973).

The *deasphalting process* is a mature process, but it is necessary to include a description of the process here so that it might be compared with new processes with options that also provide for deasphalting various feedstocks. In the deasphalting process, the feedstock is mixed with dilution solvent from the solvent accumulator and then cooled to the desired temperature before entering the extraction tower. Because of its high viscosity, the charge oil can neither be cooled easily to the required temperature nor be mixed readily with solvent in the extraction tower. By adding a relatively small portion of solvent upstream of the charge cooler (insufficient to cause phase separation), the viscosity problem is avoided.

The feedstock, with a small amount of solvent, enters the extraction tower at a point about two thirds up the column. The solvent is pumped from the accumulator, cooled, and enters near the bottom of the tower. The extraction tower is a multistage contactor, normally equipped with baffle trays. The heavy oil flows downward while the light solvent flows upward. As the extraction progresses, the desired oil goes to the solvent and the asphalt separates and moves toward the bottom. As the extracted oil and solvent rise in the tower, the temperature is increased in order to control the quality of the product by providing adequate reflux for optimum separation. The separation of the oil from the asphalt is controlled by maintaining a temperature gradient across the extraction tower and by varying the solvent/oil ratio. The tower-top temperature is regulated by adjusting the feed inlet temperature and the steam flow to the heating coils in the top of the tower. The temperature at the bottom of the tower is maintained at the desired level by the temperature of the entering solvent. The deasphalted oil-solvent mixture flows from the top of the tower under pressure control to a kettle-type evaporator heated by low-pressure steam. The vaporized solvent flows through the condenser into the solvent accumulator.

The liquid phase flows from the bottom of the evaporator, under level control, to the deasphalted oil flash tower where it is reboiled by means of a fired heater. In the flash tower, most of the remaining solvent is vaporized and flows overhead, joining the solvent from the low-pressure steam evaporator. The deasphalted oil, with relatively minor solvent, flows from the bottom of the flash tower, under level control, to a steam stripper operating at essentially atmospheric pressure. Superheated steam is introduced into the lower portion of the tower. The remaining solvent is stripped out and flows overhead with the steam through a condenser into the compressor suction drum, where the water drops out. The water flows from the bottom of the drum, under level control, to appropriate disposal.

The asphalt-solvent mixture is pressured from the extraction tower bottom to the asphalt heater and onto the asphalt flash drum, where the vaporized solvent is separated from the asphalt. The drum operates essentially at the solvent-condensing pressure so that the overhead vapors flow directly through the condenser into the solvent accumulator.

Hot asphalt with a small quantity of solvent flows from the asphalt flash drum bottom, under level control, to the asphalt stripper, which is operated at near atmospheric pressure. Superheated steam is introduced into the bottom of the stripper. The steam and solvent vapors pass overhead, join the deasphalted oil stripper overhead, and flow through the condenser into the compressor suction drum. The asphalt is pumped from the bottom of the stripper, under level control, to storage.

The yield of deasphalted oil varies with the feedstock, but the deasphalted oil does make less coke and more distillate than the feedstock. The metals content of the deasphalted oil is relatively low. The nitrogen and sulfur contents in the deasphalted oil are also related to the deasphalted oil yield.

The process parameters for a deasphalting unit must be selected with care according to the nature of the feedstock and the desired final products. The *choice of solvent* is vital to the flexibility and performance of the unit. The solvent must be suitable not only for the extraction of the desired oil fraction but also for the control of the yield and/or quality of the deasphalted oil at temperatures that are within the operating limits. If the temperature is too high (i.e., close to the critical temperature of the solvent), the operation becomes unreliable in terms of product yields and character. If the temperature is too low, the feedstock may be too viscous and have an adverse effect on the contact with the solvent in the tower. Liquid propane is by far the most selective solvent among the light hydrocarbons used for deasphalting. At temperatures ranging from 38°C to 65°C (100°F to 150°F), most hydrocarbons are soluble in propane, whereas asphaltic and resinous compounds are not. This thereby allows the rejection of these compounds, which results in a drastic reduction (relative to the feedstock) of the nitrogen content and the metals in the deasphalted oil. Although the deasphalted oil from propane deasphalting has the best quality, the yield is usually less than the yield of deasphalted oil produced using a higher-molecular-weight (higher-boiling) solvent.

The ratios of propane to oil required vary from 6 to l to 10 to 1 by volume, with the ratio occasionally being as high as 13 to –1. Since the critical temperature of propane is 97°C (206°F), this limits the extraction temperature to about 82°C (180°F). Therefore, because of the relatively low operating temperature, propane alone may not be suitable for high-viscosity feedstocks.

Iso-butane and *n*-butane are more suitable for deasphalting high-viscosity feedstocks since their critical temperatures are higher (134°C, 273°F, and 152°C, 306°F, respectively) than that of propane. Higher extraction temperatures can be used to reduce the viscosity of the heavy feed and to increase the transfer rate of oil to solvent.

Although *n*-pentane is less selective for metals and carbon residue removal, it can increase the yield of deasphalted oil from a heavy feed by a factor of two to three over propane (Speight, 2000, 2007). However, if the content of the metals and carbon residue of the pentane-deasphalted oil is too high (defined by the ensuing process), the deasphalted oil may be unsuitable as a cracking feedstock. In certain cases, the nature of the cracking catalyst may dictate that the pentane-deasphalted oil be blended with vacuum gas oil that, after further treatment such as hydrodesulfurization, produces a good cracking feedstock.

Solvent composition is an important process variable for deasphalting units. The use of a single solvent may (depending on the nature of the solvent) limit the range of feedstocks that can be processed in a deasphalting unit. When a deasphalting unit is required to handle a variety of feedstocks and/or produce various yields of deasphalted oil (as is the case in these days of variable feedstock quality), a dual solvent may be the only option to provide the desired flexibility. For example, a mixture of propane and *n*-butane might be suitable for feedstocks that vary from vacuum residua to the heavy resid to heavy gas oils that contain asphaltic materials. Adjusting the solvent composition allows the most desirable product quantity and quality within the range of temperature control.

Besides the solvent composition, the *solvent/oil ratio* also plays an important role in a deasphalting operation. Solvent/oil ratios vary considerably and are governed by feedstock characteristics and desired product qualities. For each individual feedstock, there is a minimum operable solvent/oil ratio. Generally, increasing the solvent/oil ratio almost invariably results in improving the deasphalted oil quality at a given yield, but other factors must also be taken into consideration, and (generalities aside) each plant and feedstock will have an optimum ratio.

The main consideration in the selection of the *operating temperature* is its effect on the yield of deasphalted oil. For practical applications, the lower limits of operable temperature are set by the viscosity of the oil-rich phase. When the operating temperature is near the critical temperature of the solvent, control of the extraction tower becomes difficult since the rate of change of solubility with temperature becomes very large at conditions close to that critical point. Such changes in solubility cause large amounts of oil to transfer between the solvent-rich and oil-rich phases. This, in turn, causes *flooding* and/or uncontrollable changes in product quality. In summary, the upper limits of operable temperatures must lie below the critical temperature of the solvent in order to ensure good control of the product quality and to maintain a stable condition in the extraction tower.

The *temperature gradient* across the extraction tower influences the sharpness of separation of the deasphalted oil and the asphalt because of internal reflux that occurs when the cooler oil/solvent solution in the lower section of the tower attempts to carry a large portion of oil to the top of the tower. When the oil/solvent solution reaches the steam-heated, higher-temperature area near the top of the tower, some oil of higher molecular weight in the solvent solution is rejected because the oil is less soluble in solvent at the higher temperature. The heavier oil (rejected from the solution at the top of the tower) attempts to flow downward and causes the internal reflux. Generally, the greater the temperature difference between the top and the bottom of the tower, the greater the internal reflux will be and the better the quality of the deasphalted oil will be. However, too much internal reflux can cause tower flooding and jeopardize the process.

The *process pressure* is usually not considered to be an operating variable since it must be higher than the vapor pressure of the solvent mixture at the tower-operating temperature to maintain the solvent in the liquid phase. The tower pressure is usually only subject to change when there is a need to change the solvent composition or the process temperature.

Proper *contact and distribution of the oil and solvent* in the tower are essential to the efficient operation of any deasphalting unit. In early units, mixer-settlers were used as contactors, but they proved to be less efficient than the countercurrent contacting devices. Packed towers are difficult to operate in deasphalting process because of the large differences in viscosity and density between the asphalt phase and the solvent-rich phase.

The *extraction tower* for solvent deasphalting consists of two contacting zones: (1) a rectifying zone above the oil feed and (2) a stripping zone below the oil feed. The rectifying zone contains some elements designed to promote contacting and to avoid *channeling*. Steam-heated coils are provided to raise the temperature sufficiently to induce an oil-rich reflux in the top section of the tower. The stripping zone has disengaging spaces at the top and bottom and consists of contacting elements between the oil inlet and the solvent inlet.

A *countercurrent tower* with static baffles is widely used in solvent deasphalting service. The baffles consist of fixed elements formed of expanded metal gratings in groups of two or more to provide maximum change of direction without limiting capacity. The *rotating disk contactor* has also been employed and consists of disks connected to a rotating shaft, used in place of the static baffles in the tower. The rotating element is driven by a variable-speed drive at either the top or the bottom of the column, and operating flexibility is provided by controlling the speed of the rotating element and thus the amount of mixing in the contactor.

The solvent may be separated from the deasphalted oil in several ways, such as conventional evaporation or the use of a flash tower. Irrespective of the method of solvent recovery from the deasphalted oil, it is usually most efficient to recover the solvent at a temperature close to the extraction temperature. If a higher temperature for solvent recovery is used, heat is wasted in the form of high vapor temperature. Conversely, if a lower temperature is used, the solvent must be reheated, thereby requiring additional energy input. The solvent-recovery pressure should be low enough to maintain a smooth flow under pressure from the extraction tower.

As always, the deasphalter reject (solvent asphalt) remains. It can be used (apart from its use for various types of asphalt) as feed to a partial oxidation unit to make a hydrogen-rich gas for use in hydrodesulfurization processes and hydrocracking processes. Alternatively, the asphalt may be treated in a visbreaker to reduce its viscosity, thereby minimizing the need for cutter stock to be blended with the solvent asphalt for making fuel oil. Hydrovisbreaking offers an option of converting the asphalt to feedstocks for other conversion processes.

The *deep solvent deasphalting (DSD) process* is an application of the low energy deasphalting (LEDA) process that is used to extract high-quality lubricating-oil bright stock or prepare catalytic cracking feeds, hydrocracking feeds, hydrodesulfurizer feeds, and asphalt from vacuum residua (RAROP, 1991, p. 91; Hydrocarbon Processing, 1998, p. 67).

The LEDA process uses a low-boiling hydrocarbon solvent specifically formulated to ensure the most economical deasphalting design for each operation. The DSD process can be integrated with a delayed coking operation (ASCOT process); in that case, the solvent can be a low-boiling naphtha. Low-energy deasphalting operations are usually carried out in a rotating disc contactor (RDC), which provides more extraction stages than a mixer-settler or baffle-type column. Although not essential to the process, the rotating disc contactor provides higher quality deasphalted oil at the same yield or higher yields of the same quality.

In the process, feedstock is combined with a small quantity of solvent to reduce its viscosity and cooled to a specific extraction temperature before entering the rotating disc contactor. Recovered solvent from the high-pressure and low-pressure solvent receivers are combined, adjusted to a specific temperature by the solvent heater-cooler, and injected into the bottom section of the rotating disc contactor. Solvent flows upward, extracting the paraffinic hydrocarbons from the feedstock, which is flowing downward through the rotating disc contactor. Steam coils at the top of the tower maintain the specified temperature gradient across the rotating disc contactor. The higher temperature in the top section of the rotating disc contactor results in separation of the less soluble heavier material from the deasphalted oil mix and provides internal reflux, which improves the separation. The deasphalted oil mix leaves the top of the rotating disc contactor tower. It flows to an evaporator where it is heated to vaporize a portion of the solvent. It then flows into the high-pressure flash tower; high-pressure solvent vapors are taken overhead.

The deasphalted oil mix from the bottom of this tower flows to the pressure vapor heat exchanger, where additional solvent is vaporized from the deasphalted oil mix by condensing high-pressure flash. The high-pressure solvent, totally condensed, flows to the high-pressure solvent receiver. Partially vaporized, the deasphalted oil mix flows from the pressure vapor heat exchanger to the low-pressure flash tower, where low-pressure solvent vapor is taken overhead, condensed, and collected in the low-pressure solvent receiver. The deasphalted oil mix flows down the low-pressure flash tower to the reboiler, where it is heated, and then to the deasphalted oil stripper, where the remaining solvent is stripped overhead with superheated steam. The deasphalted oil product is pumped from the stripper bottom and is cooled, if required, before flowing to battery limits.

The raffinate phase, containing asphalt and a small amount of solvent, flows from the bottom of the rotating disc contactor to the asphalt mix heater. The hot, two-phase asphalt mix from the heater is flashed in the asphalt-mix flash tower, where solvent vapor is taken overhead, condensed, and collected in the low-pressure solvent receiver. The remaining asphalt mix flows to the asphalt stripper, where the remaining solvent is stripped overhead with superheated steam. The asphalt-stripper overhead vapors are combined with the overhead vapor from the deasphalted oil stripper, condensed, and collected in the stripper drum. The asphalt product is pumped from the stripper and is cooled by generating low-pressure steam.

The *MDS Process* is a technical improvement of the solvent deasphalting process, particularly effective for upgrading heavy crude oils (Kashiwara, 1980; RAROP, 1991, p. 95). Combined with hydrodesulfurization, the process is fully applicable to the feed preparation for fluid catalytic cracking and hydrocracking. The process is capable of using a variety of feedstocks, including atmospheric and vacuum residua derived from various crude oils, oil sand, and visbreaker non-volatile products. In the process, the feed and the solvent are mixed and fed to the deasphalting tower. Deasphalting extraction proceeds in the upper half of the tower. After the removal of the asphalt, the mixture of deasphalted oil and solvent flows out of the tower through the tower top. Asphalt flows downward to come in contact with a countercurrent of rising solvent. The contact eliminates oil from the asphalt; the asphalt then accumulates on the bottom.

Deasphalted oil containing solvent is heated through a heating furnace and fed to the deasphalted oil flash tower, where most of the solvent is separated under pressure. Deasphalted oil still containing a small amount of solvent is again heated and fed to the stripper, where the remaining solvent is completely removed. Asphalt is withdrawn from the bottom of the extractor. Since this asphalt contains a small amount of solvent, it is heated through a furnace and fed to the flash tower to remove most of the solvent. Asphalt is then sent to the asphalt stripper, where the remaining portion of solvent is completely removed.

Solvent recovered from the deasphalted oil and asphalt flash towers is cooled, condensed into liquid, and sent to a solvent tank. The solvent vapor leaving both strippers is cooled to remove water and compressed for condensation. The condensed solvent is then sent to the solvent tank for further recycling.

The *Residuum Oil Supercritical Extraction (ROSE) process* is a solvent deasphalting process that consumes minimum energy and uses a super-critical solvent recovery system; the process is of value in obtaining oils for further processing (Gearhart, 1980; RAROP, 1991, p. 97; Low et al., 1995, Hydrocarbon Processing, 1996; Northrup and Sloan, 1996). In the process, the residuum is mixed with several-fold volume of a low-boiling hydrocarbon solvent and passed into the asphaltene separator vessel. Asphaltenes rejected by the solvent are separated from the bottom of the vessel and are further processed by heating and steam stripping to remove a small quantity of dissolved solvent. The solvent-free asphaltenes are sent to a section of the refinery for further processing. The main flow, solvent and extracted oil, passes overhead from the asphaltene separator through a heat exchanger and heater into the oil separator, where the extracted oil is separated without solvent vaporization. The solvent, after heat exchange, is recycled to the process. The small amount of solvent contained in the oil is removed by steam stripping, and the resulting vaporized solvent from the strippers is condensed and returned to the process. Product oil is cooled by heat exchange before being pumped to storage or further processing.

The *Solvahl process* is a solvent deasphalting process for application to vacuum residua (RAROP, 1991, p. 9). The process was developed to give maximum yields of deasphalted oil while eliminating asphaltenes and reducing metals content to a level compatible with the reliable operation of downstream units (Peries et al., 1995; Hydrocarbon Processing, 1996).

8.2 In Situ Upgrading

The potential for in situ upgrading has been recognized for several decades when it was noted that much of the oil produced by enhanced recovery processes (especially the in situ combustion processes) had some improvement in properties over the oil in situ. This is especially the case when recovering heavy oil. Any improvement in API gravity of the recovered oil versus the in situ oil can be measured in terms of money saved for refining per point of API gravity improvement.

In order the accomplish in situ upgrading, one of two requirements must be met: (1) the addition of hydrogen or (2) the deposition of thermal coke or a cokelike sediment that removes much of the

Table 8–1 Recent Process Concepts for Refining Heavy Oil

Processes	Comments
Solvent Processes	
DSD Process	Solvent chosen for yield of deasphaltened oil.
DEMEX Process	Less selective solvent than propane.
MDS Process	Solvent deasphalting and desulfurization for feedstock to catalytic cracker.
ROSE Process	Deasphaltening.
Thermal Processes	
ASCOT Process	Combination of delayed coking and deep solvent deasphalting.
CHERRY-P Process	Feedstock slurred with coal.
ET-II Process	Feedstock mixed with high-boiling recycle oil.
Catalytic Cracking Processes	
ART Process	Efficient catalyst-feedstock contact.
HOC Process	Residuum first desulfurized.
HOT Process	Steam-iron reaction to produce hydrogen in the cracker.
Hydroconversion Processes	
ABC Process	Asphaltene cracking in the presence of hydrogen.
CANMET Process	Ferric sulfate used to promote hydrogenation to eliminate coke formation.
H-OIL Process	Ebullated bed reactor under iso-thermal conditions.
HYCAR Process	Hydrovisbreaking to control flocculation.
LC-FINING Process	Expanded bed reactor at near iso-thermal conditions.
MICROCAT-RC Process	Metal sulfide catalyst particles in a carbonaceous matrix.
BOC Process	Temperature control and conversion limited to avoid coking.
RHC Process	Trickle-flow reactor with multicatalyst system.

asphaltene and resin constituents from the oil. Since hydrogen addition must be used during surface upgrading in order to stabilize the upgraded bitumen and saleable products, the cost of partial upgrading can be offset by the reduced hydrogen requirements in a surface upgrader.

The most promising of enhanced recovery methods for partial upgrading during recovery are the combustion methods. The concept of any combustion technology requires that the oil be partially combusted and that thermal decomposition occur to other parts of the oil. This is sufficient to cause irreversible chemical and physical changes to the oil to the extent that the product is markedly different from the oil in place. Recognition of this phenomenon is essential before combustion technologies are applied to oil recovery.

Although this improvement in properties may not appear to be too drastic, nevertheless, it usually is sufficient to have major advantages for refinery operators. Any incremental increase in the units of the hydrogen/carbon ratio can save amounts of costly hydrogen during upgrading. The same principles are also operative for reductions in the nitrogen, sulfur, and oxygen contents. This latter occurrence also improves catalyst life and activity as well as reduces the metals content.

8.2.1 Solvent-Based Processes

The application of light hydrocarbon solvents to reduce or eliminate natural gas for steam generation has received significant recent interest. These light hydrocarbons also have a natural tendency to cause asphaltene constituents to separate, and therefore offer some promise for in situ upgrading. VAPEX is the most advanced process in this area.

Physical and chemical separations into fractions that might lead to more targeted process steps, including more efficiently targeted hydrogen addition. There may be some overlap here with demetallization.

An extension of solvent recovery is the combined use of solvents and thermal stimulation to achieve some degree of in situ upgrading. In all cases, a minor factor is a degree of upgrading that may occur in the new recovery methods. Potentially major factors are the likely conversion of bitumen-based residues in future for energy, power, and hydrogen at production or upgrading stages and the possible application of mild thermal upgrading in situ of field upgrading to reduce dependence on diluent for transport to distant refineries. The potential move to *less severe* primary upgrading would place more emphasis on *conversion* at the secondary stage as well as heteroatom removal. The desire to reduce overall hydrogen consumption will place emphasis on lower light by-product production and targeted hydrogen addition to synthetic crude oil.

8.2.2 Bulk Thermal Processes

The mobilization of heavy oil in the reservoir by partial combustion is not a new idea, but it still has many hurdles to overcome before it can be considered close to commercial. However, the product oil is likely less viscous. In situ recovery processes (although less efficient in terms of bitumen recovery relative to mining operations) may also have the added benefit of *leaving* some of the more obnoxious constituents (from the processing objective) in the ground.

The techniques involve methods such as the establishment of hydrovisbreaking in situ, for example, by injection of superheated steam and hot hydrogen into the formation. The injected fluids initiate hydrovisbreaking, producing a partially upgraded lighter product (that may have solvent properties) and driving the oil in the formation to the producing wells. In some process schemes, specialized downhole equipment is used to generate superheated steam and heat the injected hydrogen. However, control of in situ combustion processes have been difficult. The THAI process offers some relief for the potential for an out-of-control combustion reaction, but it should not be construed as an easy task.

Visbreaking, its variants, and the recently demonstrated ORMAT process are examples of bulk thermal processes that convert residues without progressing all the way to solid coke. These processes have significant potential integrated with deasphalting to produce residues of varying yields on bitumen to meet future alternative-energy and hydrogen-production needs.

In situ combustion has long been used as an enhanced oil recovery method. For heavy oils, numerous field observations have shown upgrading of 2° to 6° API for heavy oils undergoing combustion (Ramey et al., 1992). During in situ combustion of heavy oils, temperatures of up to 700°C can be observed at the combustion front. Some processes offer the potential for partial upgrading during recovery, and many typically involve in situ hydrogenation of the oil or bitumen (Hamrick and Rose, 1977; Gregoli 1985; Hewgill and Kalfayan, 1992; Gregoli et al., 2000; Graue, 2001).

In situ combustion is the injection of an oxidizing gas (air or oxygen-enriched air) to generate heat by burning a portion of the oil (Chapter 5). Most of the oil is driven toward the producers by a combination of gas drive (from the combustion gases) and steam and water drive. This process is also called fireflooding to describe the movement of the burning front inside the reservoir. Based on the respective directions of front propagation and air flow, the process can be forward, when the combustion front advances in the same direction as the air flow, or reverse, when the front moves against the air flow.

Forward combustion can be further characterized as "dry" when only air or enriched air is injected or "wet" when air and water are co-injected. In the process, air is injected in the target formation for a short time, usually a few days to a few weeks, and the oil in the formation is ignited. Ignition can be induced using downhole gas burners, electrical heaters, and/or the injection of pyrophoric agents (not recommended) or steam. In some cases, autoignition occurs when the reservoir temperature is fairly high and the oil reasonably reactive. This often happens for California oils.

After ignition, the combustion front is propagated by a continuous flow of air. As the front progresses into the reservoir, several zones can be found between the injector and the producer as a result of heat, mass transport, and the chemical reactions occurring in the process. The burned zone is the volume already burned. This zone is filled with air and may contain small amounts of residual unburned organic solids. As it has been subjected to high temperatures, mineral alterations are possible. Because of the continuous air flow from the injector to the burned zone, temperature increases from injected-air temperature at the injector to near combustion-front temperature near the combustion front. There is no oil left in this zone.

The combustion front is the highest temperature zone. It is very thin, often no more than several inches thick. It is in that region that oxygen combines with the fuel, and high temperature oxidation occurs. The products of the burning reactions are water and carbon oxides. The fuel is often misnamed coke. In fact, it is not pure carbon but a hydrocarbon with H/C atomic ratios ranging from about 1 to 2.0. This fuel is formed in the thermal cracking zone just ahead of the front and is the product of cracking and pyrolysis, which is deposited on the rock matrix. The amount of fuel burned is an important parameter because it determines how much air must be injected to burn a certain volume of reservoir.

Chemical reactions are of two main categories: (1) oxidation, which occurs in the presence of oxygen, and (2) pyrolysis, which is caused mainly by elevated temperatures. In general, at low temperature, oxygen combines with the oil to form oxidized hydrocarbons such as peroxides, alcohols, or ketones. This generally increases the oil viscosity but could increase oil reactivity at a higher temperature. When oxygen contacts the oil at a higher temperature, combustion occurs, resulting in the production of water and carbon oxides. Of all the reactions that can occur during in situ combustion, only low-temperature oxidation can increase the viscosity of the oil. If the fire flood is conducted properly, low-temperature oxidations are minimized because most of the oxygen injected is consumed at the burning front.

Distillation allows the transport and production of the light fractions of the oil, leaving behind the heavy ends. These heavy ends often contain the majority of the undesirable compounds, which may contain sulfur or metals.

Forward in situ combustion by itself is already an effective in situ upgrading method and, in field tests, has made improvements in the API gravity by as much as 6° (Ramey et al., 1992).

Another possible in situ upgrading technique would involve a combination of solvent injection and combustion. Cyclic injection of solvents, either gas or liquid, would be followed by in situ combustion of a small part of the reservoir to increase the temperature near the well and to clean the well bore region of all the residues left by the solvents. Alternate slugs of solvent and air would be injected, and production would occur after each solvent slug injection and after each combustion period. The process could be repeated until an economic limit were reached. One important fact to note is that both solvent injection and in situ combustion have been proven to be effective in a variety of reservoirs, but the combination of the two methods has never been tried.

The most significant effect would be the precipitation and/or deposition of heavy hydrocarbons such as asphaltenes or paraffins. The produced oil would be expected to be slightly upgraded by the solvent cycle. Unlike the classic well-to-well in situ combustion, near–well bore conditions could be improved by burning the solid residues left after the solvent cycle. The benefits of using combustion at this stage would be expected to include: (1) productivity improvement through removal of the heavy ends left from the solvent cycle, (2) possible deactivation of the clays near the well bore due to the high temperature of the combustion, and (3) reduced viscosity of the oil due to the temperature increase.

Downhole upgrading of virgin Athabasca tar sand bitumen was investigated in a series of experiments using the THAI process, which uses combinations of vertical injection wells and horizontal producer wells arranged in a direct, or staggered line drive. Downhole upgrading of the bitumen was significant, with the API gravity of the produced oil increasing by an average of 8° API, compared to the original bitumen. The produced oil viscosity was also dramatically reduced. SARA analysis was used to assess the quality of the produced oil, and it showed that the saturate fraction of the bitumen was increased from approximately 16% by weight to 72% by weight.

The THAI process could well have a wider range of application than SAGD, but in any case, a detailed knowledge of the reservoir is essential. SAGD generally works best in relatively thick (40m) homogeneous pay zones. It is possible that the THAI process would be effective down to about 6m thickness, as is common in many Saskatchewan heavy oil pools.

The CAPRI process involves the addition of a gravel-packed catalyst, as used in a conventional refinery, between the tubing and the horizontal well bore. Test results have shown the technique to add 6° to 8° API on top of the THAI in situ upgrades. Based on these data, the combination could deliver in situ upgrading to above the 22° API requirement for produced fluids that can be transported by pipeline without diluent. This also represents major savings in surface-upgrading and refining costs.

The benefits of the introduction of hydrogen during in situ upgrading offer much promise. The possible application of in situ upgrading methods for selective separation of the metal constituents is an obvious benefit. For example, partial oxidation in the presence of steam may produce hydrogen for immediate pickup and result in integrated recovery and significant upgrading.

8.3 References

Aiba, T., Kaji, H., Suzuki, T., and Wakamatsu, T. 1981. Chemical Engineering Progress. February:37.

Bartholic, D.B., Center, A.M., Christian, B.R., and Suchanek, A.J. 1992. Petroleum Processing Handbook. Ed. J.J. McKetta. Marcel Dekker Inc., New York. p. 108.

Bonilla, J. 1985. Energy Progress. December:5.

Bonilla, J., and Elliott, J.D. 1987. Asphalt Coking Method. United States Patent 4,686,027.

Chen, R. 1993. Preprints. Oil Sands Our Petroleum Future. Alberta Oil Sands Technology and Research Authority, Edmonton, Alberta, Canada. p. 287.

Dickenson, R.L., Biasca, F.E., Schulman, B.L., and Johnson, H.E. 1997. Hydrocarbon Processing. 76(2):57.

Ditman, J.G. 1973. Hydrocarbon Processing. 52(5):110.

Fattouh, B. 2006. OPEC's Discounts on Heavy Crude Oil: Is a New Policy Instrument Taking Shape? Oxford Energy Comment. Oxford Institute for Energy Studies, Oxford, England.

Feldman, J.A., Lutter, B.E., and Hair, R.L. 1992. Petroleum Processing Handbook. Ed. J.J. McKetta. Marcel Dekker Inc., New York. p. 480.

Finneran, J.A. 1974. Oil and Gas Journal. January 14. (2):72.

Gearhart, J.A. 1980. Hydrocarbon Processing. 59(5):150.

Graue, D.J. 2001. Upgrading and Recovering Heavy Crude Oils and Natural Bitumens by In Situ Hydrovisbreaking. United States Patent 6,328,104.

Green, P. M., and Center, A. M. 1985. Proceedings, 1219. Third UNITAR/UNDP International Conference on Heavy Crude and Tar Sands. Long Beach, California.

Gregoli, A.A. 1985. Method of In Situ Hydrogenation of Carbonaceous Material. United States Patent 4,501,445.

Gregoli, A.A., and Rimmer, D.P. 2000. Production of Synthetic Crude oil from Heavy Hydrocarbons Recovered by In Situ Hydrovisbreaking. United States Patent 6,016.868.

Gunter, W., Mitchell, R., Potter, I., Lakeman, B., and Wong, S. 2005. The CANiCAP Program. Alberta Research Council, Edmonton, Alberta, Canada. http://www.arc.ab.ca/documents/CANiCAP%20Final%20Report.pdf

Gussow, S., and Kramer, R. 1990. Annual Meeting. National Petroleum Refiners Association. March.

Hamrick, J.T., and Rose, L.C. 1977. In Situ Hydrogenation of Hydrocarbons in Underground Formations. United States Patent 4,050,525.

Hedrick, B.W., Seibert, K.D., and Crewe, C. 2006. A New Approach to Heavy Oil and Bitumen Upgrading. Report No. AM-06-29. UOP LLC, Des Plaines, Illinois.

Hewgill, G.S., and Kalfayan, L.J. 1992. Enhanced Oil Recovery Technique Using Hydrogen Precursors. United States Patent 5,105,887.

Hydrocarbon Processing. 1996. 75(11):89 et seq.

Hydrocarbon Processing. 1998. 77(11):53 et seq.

Johnson, T.E. 1982. Oil and Gas Journal. March. 22.

Kashiwara, H. 1980. Kagaku Kogaku. (7):44.

Khan, M. R., and Patmore, D.J. 1998. Heavy Oil Upgrading Processes. Chapter 6. In Petroleum Chemistry and Refining, ed. J.G. Speight. Taylor & Francis, Washington, DC.

Kuwahara, I. 1987. Kagaku Kogaku. 51:1.

LePage, J. F., Morel, F., Trassard, A. M., and Bousquet, J. 1987. Preprints Division of Fuel Chemistry. 32:470.

Logwinuk, A.K., and Caldwell, D.L. 1983. Annual Meeting, National Petroleum Refiners Association. March.

Low, J. Y., Hood, R. L., and Lynch, K. Z. 1995. Preprints Division of Petroleum Chemistry American Chemical Society. 40:780.

Maples, R.E. 2000. Petroleum Refinery Process Economics. 2nd Edition. PennWell Corporation, Tulsa, Oklahoma.

Marzin, R., Pereira, P., McGrath, M.J., Feintuch, H.M., and Thompson, G. 1998. Oil and Gas Journal. 97(44):79.

Miyauchi, T., Furusaki, S., and Morooka, Y. 1981. Chapter 11. In Advances in Chemical Engineering. Academic Press Inc., New York.

Miyauchi, T., and Ikeda, Y. 1988. Process for Thermal Cracking of Heavy Oil. United States Patent 4,772,378.

Moschopedis, S.E., Ozum, B., and Speight, J.G. 1998. Upgrading Heavy Oils. Reviews in Process Chemistry and Engineering. 1:201.

Murphy, J.R., and Treese, S.A. 1979. Oil and Gas Journal. June 25:135.

Northrup, A. H., and Sloan, H. D. 1996. Paper AM-96-55. Annual Meeting, National Petroleum Refiners Association. Houston, Texas.

Ozaki, Y.1982. Proceedings. 32nd Annual Conference, Canadian Society for Chemical Engineering. Vancouver, Canada.

Peries, J. P., Billon, A., Hennico, A., Morrison, E., and Morel, F., 1995. Proceedings. 6th UNITAR International Conference on Heavy Crude and Tar Sand. 2:229.

Pratts, M. 1986. Thermal Recovery. Volume 7. Society of Petroleum Engineers, New York.

Ramey, H.J., Jr., Stamp, V.V., and Pebdani, F.N., 1992. Case History of South Belridge, California, In Situ Combustion Oil Recovery. Paper No. SPE 24200. Proceedings. 9th Society of Petroleum Engineers/US Department of Energy Enhanced Oil Recovery Symposium. Tulsa, Oklahoma. April 21–24.

Research Association for Residual Oil Processing. 1991. RAROP Heavy Oil Processing Handbook. Ministry of Trade and International Industry (MITI), Tokyo, Japan.

Reynolds, B.E., Brown, E.C., and Silverman, M.A. 1992. Hydrocarbon Processing. 71(4):43.

Richardson, W.C., Fontaine, M.F., and Haynes, S. 1992. Paper No. SPE 24033. Western Regional Meeting. Bakersfield, California, March 30–April 1.

Speight, J.G. 1990. Fuel Science and Technology Handbook. Ed. J.G. Speight. Marcel Dekker Inc., New York. Chapters 12–16.

Speight, J.G. 2000. The Desulfurization of Heavy Oils and Residua. 2nd Edition. Marcel Dekker Inc., New York.

Speight, J.G. 2007. The Chemistry and Technology of Petroleum. 4th Edition. CRC-Taylor and Francis Group, Boca Raton, Florida.

Speight, J.G., and Ozum, B. 2002. Petroleum Refining Processes. Marcel Dekker Inc., New York.

Takeuchi, C. 1982. Proceedings. 2nd International Conference on the Future of Heavy Crude and Tar Sands. Caracas, Venezuela.

Thomas, K.P., Barbour, R.V., Branthaver, J.F., and Dorrence, S.M. 1983. Fuel. 62:438.

Ueda, K. 1976. Journal of the Japanese Petroleum Institute. 19(5):417.

Ueda, H. 1978. Journal of the Fuel Society of Japan. 57:963.

Washimi, K. 1989. Hydrocarbon Processing. 68(9):69.

Washimi, K., and Hamamura, M. 1993. Preprints. Oil Sands Our Petroleum Future. Alberta Oil Sands Technology and Research Authority, Edmonton, Alberta Canada. p. 283.

Watari, R., Shoji, Y., Ishikawa, T., Hirotani, H., and Takeuchi, T. 1987. Paper AM-87-43. Annual Meeting, National Petroleum Refiners Association. San Antonio, Texas.

Zou, J., Gray, M.R., and Thiel, J. 1989. AOSTRA J. Res. 5:75

APPENDIX A

CONVERSION FACTORS

1 acre = 43,560 sq ft

1 acre foot = 7758.0 bbl

1 atmosphere = 760 mm Hg = 14.696 psia = 29.91 in. Hg

1 atmosphere = 1.0133 bars = 33.899 ft. H_2O

1 barrel (oil) = 42 gal = 5.6146 cu ft

1 barrel (water) = 350 lb at 60°F

1 barrel per day = 1.84 cu cm per second

1 Btu = 778.26 ft-lb

1 centipoise × 2.42 = Ib mass/(ft) (hr), viscosity

1 centipoise × 0.000672 = Ib mass/(ft) (sec), viscosity

1 cubic foot = 28,317 cu cm = 7.4805 gal

Water density at 60°F = 0.999 gram/cu cm = 62.367 lb/cu ft = 8.337 lb/gal

1 gallon = 231 cu in. = 3,785.4 cu cm = 0.13368 cu ft

1 horsepower-hour – 0.7457 kwhr = 2544.5 Btu

1 horsepower = 550 ft-lb/sec = 745.7 watts

1 inch = 2.54 cm

1 meter = 100 cm = 1,000 mm = 10 microns = 10 angstroms (A)

1 ounce = 28.35 grams

1 pound = 453.59 grams = 7,000 grains

1 square mile = 640 acres

GLOSSARY

1P reserve A proved reserve.

2P reserves The total of proved plus probable reserves.

3P reserves The total of proved reserves plus probable reserves plus possible reserves.

Abandonment pressure A direct function of the economic premises, the static bottom pressure at which the revenues obtained from the sales of the hydrocarbons produced are equal to the well's operation costs.

Absolute permeability Ability of a rock to conduct a fluid when only one fluid is present in the pores of the rock.

Acid deposition (acid rain) A form of pollution depletion in which pollutants, such as nitrogen oxides and sulfur oxides, are transferred from the atmosphere to soil or water. It is often referred to as atmospheric self-cleaning. The pollutants usually arise from the use of fossil fuels.

Acidizing A technique for improving the permeability of a reservoir by injecting acid.

Acid number A measure of the reactivity of petroleum with a caustic solution, given in terms of milligrams of potassium hydroxide that are neutralized by one gram of petroleum.

Acid rain The precipitation phenomenon that incorporates anthropogenic acids and other acidic chemicals from the atmosphere to the land and water (see also acid deposition).

Acoustic log See sonic log.

Acre-foot A measure of bulk rock volume where the area is one acre and the thickness is one foot.

Additions The reserve provided by the exploratory activity. It consists of the discoveries and delimitations in a field during the study period.

After flow Flow from the reservoir into the well bore that continues for a period after the well has been shut in. After-flow can complicate the analysis of a pressure transient test.

Air injection An oil recovery technique using air to force oil from the reservoir into the well bore.

Alkaline flooding See EOR process.

American Society for Testing and Materials (ASTM) The official organization in the United States for designing standard tests for petroleum and other industrial products.

Anticline Structural configuration of a package of folding rocks in which the rocks are tilted in different directions from the crest.

API gravity A measure of the *lightness* or *heaviness* of petroleum, that is related to density and specific gravity.

°API = (141.5/sp gr @ 60°F) – 131.5.

Apparent viscosity The viscosity of a fluid, or several fluids flowing simultaneously, measured in a porous medium (rock) and subject to both viscosity and permeability effects; it is also called effective viscosity.

Aquifer A subsurface rock interval that will produce water; it is often the underlay of a petroleum reservoir.

Areal sweep efficiency (horizontal sweep efficiency) The fraction of the flood pattern area that is effectively swept by the injected fluids.

Artificial production system Any of the techniques used to extract petroleum from the producing formation to the surface when the reservoir pressure is insufficient to raise the oil naturally to the surface.

Associated gas in solution or dissolved Natural gas dissolved in the crude oil of the reservoir, under the prevailing pressure and temperature conditions.

Associated gas Natural gas that is in contact with and/or dissolved in the crude oil of the reservoir. It may be classified as gas cap (free gas) or gas in solution (dissolved gas).

Asphaltene (asphaltenes) The brown-to-black powdery material produced by treatment of petroleum, petroleum residua, or bituminous materials with a low-boiling liquid hydrocarbon, e.g. pentane or heptane; it is soluble in benzene (and other aromatic solvents), carbon disulfide, and chloroform (or other chlorinated hydrocarbon solvents).

Asphaltene association factor The number of individual asphaltene species that associate in nonpolar solvents as measured by molecular weight methods; also, the molecular weight of asphaltenes in toluene divided by the molecular weight in a polar non-associating solvent, such as dichlorobenzene, pyridine, or nitrobenzene.

Barrel The unit of measurement of liquids in the petroleum industry, equivalent to 42 U.S. standard gallons or 33.6 imperial gallons.

Basement The foot or base of a sedimentary sequence composed of igneous or metamorphic rocks.

Basic nitrogen Nitrogen in petroleum that occurs in pyridine form.

Basic sediment and water (bs&w, bsw) The material that collects in the bottom of storage tanks, usually composed of oil, water, and foreign matter; it is also called bottoms or bottom settlings.

Basin A receptacle in which a sedimentary column is deposited that shares a common tectonic history at various stratigraphic levels.

Baumé gravity The specific gravity of liquids expressed as degrees on the Baumé (°Bé) scale. For liquids lighter than water, Sp gr 60°F = 140/(130 + °Bé). For liquids heavier than water, Sp gr 60°F = 145/(145 – °Bé).

Bell cap A hemispherical or triangular cover placed over the riser in a (distillation) tower to direct the vapors through the liquid layer on the tray. See also bubble cap.

Billion $1 \times 10^{\circ}$

Bitumen The portion of petroleum that exists in the reservoirs in a semi-solid or solid phase. In its natural state, it generally contains sulfur, metals, and other non-hydrocarbon compounds. Natural bitumen has a viscosity of more than 10,000 centipoises, measured at the original temperature of the reservoir, at atmospheric pressure, and gas-free. It frequently requires treatment before being refined.

Bituminous Containing bitumen or constituting the source of bitumen.

Bituminous rock See Bituminous sand.

Bituminous sand A formation in which the bituminous material (see bitumen) is found filling in veins and fissures in fractured rock or impregnating relatively shallow sand, sandstone, and limestone strata; also a sandstone reservoir that is impregnated with a heavy, viscous, black, petroleum-like material that cannot be retrieved through a well by conventional production techniques.

Boiling point A characteristic physical property of a liquid at which the vapor pressure is equal to that of the atmosphere and the liquid is converted to a gas.

Boiling range The range of temperature, usually determined at atmospheric pressure in a standard laboratory apparatus, over which the distillation of oil commences, proceeds, and finishes.

British thermal unit (Btu) The energy required to raise the temperature of one pound of water one degree Fahrenheit.

Btu See British thermal unit.

Calorific equivalence of dry gas to liquid factor (cedglf) The factor used to relate dry gas to its liquid equivalent. It is obtained from the molar composition of the reservoir gas, considering the unit heat value of each component and the heat value of the equivalence liquid.

Capillary forces Interfacial forces between immiscible fluid phases, resulting in pressure differences between the two phases.

Capillary number (N_c) The ratio of viscous forces to capillary forces, equal to viscosity times velocity divided by interfacial tension.

Capillary pressure A force per area unit resulting from the surface forces to the interface between two fluids.

Catagenesis The alteration of organic matter during the formation of petroleum that may involve temperatures in the range 50°C (120°F) to 200°C (390°F). See also diagenesis and metagenesis.

Code of Federal Regulations (CFR) Title 40 (40 CFR) contains the regulations for protection of the environment.

Chemical flooding See enhanced oil recovery process.

Clastic Composed of pieces of pre-existing rock.

Cloud point The temperature at which paraffin wax or other solid substances begin to crystallize or separate from the solution, imparting a cloudy appearance to the oil when the oil is chilled under prescribed conditions.

Coal An organic rock.

Coal tar Tar produced from coal.

Coal tar pitch The specific name for the pitch produced from coal.

COFCAW An enhanced oil recovery process that combines forward combustion and waterflooding.

Cogeneration An energy conversion method by which electrical energy is produced along with steam generated for enhanced oil recovery use.

Cold production The use of operation and specialized exploitation techniques in order to rapidly produce heavy oils without using thermal recovery methods.

Completion interval The portion of the reservoir formation placed in fluid communication with the well by selectively perforating the well bore casing.

Complex A series of fields sharing common surface facilities.

Composition The general chemical makeup of petroleum.

Compressor A device installed in the gas pipeline to raise the pressure and guarantee the fluid flow through the pipeline.

Condensate A mixture of light hydrocarbon liquids obtained by condensation of hydrocarbon vapors. It is predominately butane, propane, and pentane, with some heavier hydrocarbons and relatively little methane or ethane. See also natural gas liquids.

Condensate recovery factor (crf) The factor used to obtain liquid fractions recovered from natural gas in the surface distribution and transportation facilities. It is obtained from the gas and condensate handling statistics of the last annual period in the area corresponding to the field being studied.

Conductivity A measure of the ease of flow through a fracture, perforation, or pipe.

Conformance The uniformity with which a volume of the reservoir is swept by injection fluids in area and vertical directions.

Contingent resource The amounts of hydrocarbons estimated at a given date that are potentially recoverable from known accumulations but are not considered commercially recoverable under the economic evaluation conditions corresponding to that date.

Conventional limit The reservoir limit established according to the degree of knowledge of (or research into) the geological, geophysical, or engineering data available.

Conventional recovery Primary and/or secondary recovery.

Core A cylindrical rock sample taken from a formation when drilling, used to determine the rock's permeability, porosity, hydrocarbon saturation, and other productivity-associated properties.

Core floods Laboratory flow tests through samples (cores) of porous rock.

Centipoise (cp) A unit of viscosity.

Cracking Heat and pressure procedures that transform the hydrocarbons with a high molecular weight and boiling point to hydrocarbons with a lower molecular weight and boiling point.

Cracking temperature The temperature (350°C; 660°F) at which the rate of thermal decomposition of petroleum constituents becomes significant.

Craig-Geffen-Morse method A method for predicting oil recovery by waterflood.

Crude assay A procedure for determining the general distillation characteristics (e.g., distillation profile) and other quality information of crude oil.

Crude oil See petroleum.

Cryogenic plant A processing plant capable of producing liquid natural gas products, including ethane, at very low operating temperatures.

Cryogenics The study, production, and use of low temperatures.

Cut point The boiling-temperature division between distillation fractions of petroleum.

Cyclic steam injection The alternating injection of steam and production of oil with condensed steam from the same well or wells.

Deasphaltened oil The fraction of petroleum after the asphaltene constituents have been removed.

Deasphaltening The removal of a solid powdery asphaltene fraction from petroleum by the addition of low-boiling liquid hydrocarbons such as *n*-pentane or *n*-heptane under ambient conditions.

Deasphalting The removal of the asphaltene fraction from petroleum by the addition of a low-boiling hydrocarbon liquid such as *n*-pentane or *n*-heptane; more correctly it is the removal of asphalt (tacky, semi-solid) from petroleum (as occurs in a refinery asphalt plant) by the addition of liquid propane or liquid butane under pressure.

Delimitation An exploration activity that increases or decreases reserves by means of drilling delimiting wells.

Density The mass (or weight) of a unit volume of any substance at a specified temperature. See also specific gravity.

Desalting The removal of mineral salts (mostly chlorides) from crude oils.

Developed proved area The plant projection of the extension drained by the wells of a producing reservoir.

Developed proved reserves Reserves that are expected to be recovered in existing wells, including reserves behind pipe, which may be recovered with the current infrastructure through additional work and with moderate investment costs. Reserves associated with secondary and/or enhanced recovery processes will be considered as developed when the infrastructure required for the process has been installed or when the costs required for such are lower. This category includes reserves in completed intervals that have been opened at the time when the estimation is made but that have not started flowing due to market conditions, connection problems, or mechanical problems and whose rehabilitation cost is relatively low.

Development well A well drilled in a proved area in order to produce hydrocarbons.

Development Activity that increases or decreases reserves by means of drilling exploitation wells.

Dew point pressure The pressure at which the first drop of liquid is formed, when it goes from the vapor phase to the two-phase region.

Diagenesis The concurrent and consecutive chemical reactions that commence the alteration of organic matter (at temperatures up to 50°C [120°F]) and ultimately result in the formation of petroleum from the marine sediment. See also catagenesis and metagenesis.

Diagenetic rock Rock formed by conversion (through pressure or chemical reaction) from a rock, e.g., sandstone is a diagenetic.

Differential-strain analysis The measurement of thermal stress relaxation in a recently cut well.

Dispersion A measure of the convective fluids due to flow in a reservoir.

Discovered resource The volume of hydrocarbons tested through wells drilled.

Discovery The incorporation of reserves attributable to drilling exploratory wells that test hydrocarbon-producing formations.

Displacement efficiency The ratio of the amount of oil moved from the zone swept by the recovery process to the amount of oil present in the zone prior to the start of the process.

Dissolved gas-oil ratio The ratio of the volume of gas dissolved in oil compared to the volume of oil containing gas. The ratio may be original (Rsi) or instantaneous (Rs).

Distribution coefficient A coefficient that describes the distribution of a chemical in reservoir fluids, usually defined as the equilibrium concentrations in the aqueous phases.

Dome A geological structure with a semi-spherical shape or relief.

Downhole steam generator A generator installed downhole in an oil well to which oxygen-rich air, fuel, and water are supplied for the purposes of generating steam in the reservoir. Its major advantage over a surface steam-generating facility is that the losses to the well bore and surrounding formation are eliminated.

Drainage radius The distance from which fluids flow to the well, that is, the distance reached by the influence of disturbances caused by pressure drops.

Drill stem test (formation test) The conventional formation test method.

Dry gas equivalent to liquid (DGEL) The volume of crude oil that, because of its heat rate, is equivalent to the volume of dry gas.

Dry gas Natural gas containing negligible amounts of hydrocarbons heavier than methane. Dry gas is also obtained from the processing complexes.

Dykstra-Parsons coefficient An index of reservoir heterogeneity arising from permeability variation and stratification.

Economic limit The point at which the revenues obtained from the sale of hydrocarbons match the costs incurred in its exploitation.

Economic reserves The accumulated production that is obtained from a production forecast in which economic criteria are applied.

Effective permeability A relative measure of the conductivity of a porous medium for a fluid when the medium is saturated with more than one fluid. This implies that the effective permeability is a property associated with each reservoir flow, for example, gas, oil and water. A fundamental principle is that the total of the effective permeability is less than or equal to the absolute permeability.

Effective porosity A fraction that is obtained by dividing the total volume of communicated pores by the total rock volume.

Effective viscosity See apparent viscosity.

Enhanced oil recovery (EOR) Petroleum recovery following recovery by conventional (i.e., primary and/or secondary) methods.

Enhanced oil recovery (EOR) process A method for recovering additional oil from a petroleum reservoir beyond that economically recoverable by conventional primary and secondary recovery methods. EOR methods are usually divided into three main categories: (1) *chemical flooding*, injection of water with added chemicals into a petroleum reservoir. The chemical processes include surfactant flooding, polymer flooding, and alkaline flooding; (2) *miscible flooding*, injection into a petroleum reservoir of a material that is miscible, or can become miscible, with the oil in the reservoir. Carbon dioxide, hydrocarbons, and nitrogen are used; (3) *thermal recovery*, injection of steam into a petroleum reservoir or propagation of a combustion zone through a reservoir by air or oxygen-enriched air injection. The thermal processes include steam drive, cyclic steam injection, and in situ combustion.

Evaporites Sedimentary formations consisting primarily of salt, anhydrite, or gypsum, as a result of evaporation in coastal waters.

Expanding clays Clays that expand or swell on contact with water, e.g., montmorillonite.

Exploratory well A well that is drilled without detailed knowledge of the underlying rock structure in order to find hydrocarbons whose exploitation is economically profitable.

Extra heavy oil Crude oil with relatively high fractions of heavy components, high specific gravity (low API density), and high viscosity at reservoir conditions. The production of this kind of oil generally implies difficulties in extraction and high costs. Thermal recovery methods are the most common form of commercially exploiting this kind of oil.

Facies One or more layers of rock that differs from other layers in composition, age, or content.

FAST Fracture assisted steam flood technology.

Fault A fractured surface of geological strata along which there has been differential movement.

Field scale The application of enhanced oil recovery processes to a significant portion of a field.

Fingering The formation of finger-shaped irregularities at the leading edge of a displacing fluid in a porous medium that move out ahead of the main body of fluid.

Five-spot An arrangement or pattern of wells with four injection wells at the comers of a square and a producing well in the center of the square.

Flood, flooding The process of displacing petroleum from a reservoir by the injection of fluids.

Fluid A reservoir gas or liquid.

Fluid saturation The portion of the pore space occupied by a specific fluid; oil, gas and water may exist.

Formation An interval of rock with distinguishable geologic characteristics.

Formation resistance factor (F) The ratio between the resistance of rock saturated 100% with brine divided by the resistance of the saturating water.

Formation volume factor (B) The factor that relates the volume unit of the fluid in the reservoir with the surface volume.

Fossil fuel resources A gaseous, liquid, or solid fuel material formed in the ground by chemical and physical changes (diagenesis) in plant and animal residues over geological time; these include natural gas, petroleum, coal, and oil shale.

Fractional composition The composition of petroleum as determined by fractionation (separation) methods.

Free associated gas Natural gas that overlies and is in contact with the crude oil of the reservoir. It may be gas cap.

Gas cap A part of a hydrocarbon reservoir at the top that will produce only gas.

Gas compressibility ratio (Z) The ratio between an actual gas volume and an ideal gas volume. This is an adimensional amount that usually varies between 0.7 and 1.2.

Gas lift An artificial production system that is used to raise the well fluid by injecting gas down the well through tubing or through the tubing-casing annulus.

Gas-oil ratio (GOR) The ratio of reservoir gas production to oil production, measured at atmospheric pressure.

Geological province A region of large dimensions characterized by similar geological and development histories.

Graben A dip or depression formed by tectonic processes, limited by normal-type faults.

Gravitational segregation A reservoir driving mechanism in which the fluids tend to separate according to their specific gravities. For example, since oil is heavier than water, it tends to move towards the lower part of the reservoir in a water-injection project.

Gravity See API gravity.

Gravity drainage The movement of oil in a reservoir that results from the force of gravity.

Gravity segregation The partial separation of fluids in a reservoir caused by the gravity force acting on differences in density.

Gravity-stable displacement The displacement of oil from a reservoir by a fluid of a different density, where the density difference is utilized to prevent gravity segregation of the injected fluid.

Handling efficiency shrinkage factor (hesf) A fraction of natural gas that is derived from considering self-consumption and the lack of capacity to handle such. It is obtained from the gas-handling statistics of the final period in the area corresponding to the field being studied.

HCPV Hydrocarbon pore volume.

Hearn method A method used in reservoir simulation for calculating a pseudo relative permeability curve that reflects reservoir stratification.

Heat value The amount of heat released per unit of mass or per unit of volume when a substance is completely burned. The heat power of solid and liquid fuels is expressed in calories per gram or in Btu per pound. For gases, this parameter is generally expressed in kilocalories per cubic meter or in Btu per cubic foot.

Heavy oil Petroleum having an API gravity of less than 20°.

Heavy petroleum See heavy oil.

Heteroatom compounds Chemical compounds that contain nitrogen and/or oxygen and/or sulfur and /or metals bound within their molecular structures.

Heterogeneity The lack of uniformity in reservoir properties such as permeability.

Higgins-Leighton model A stream-tube computer model used to simulate waterflood.

Horst Rock of the earth's crust rising between two faults; it is the opposite of a graben.

Hot production The optimum production of heavy oils through use of enhanced thermal recovery methods.

Huff-and-puff A cyclic enhanced oil recovery method in which steam or gas is injected into a production well; after a short shut-in period, oil and the injected fluid are produced through the same well.

Hydration The association of molecules of water with a substance.

Hydraulic fracturing The opening of fractures in a reservoir by high-pressure, high-volume injection of liquids through an injection well.

Hydrocarbon index An amount of hydrocarbons contained in a reservoir per unit area.

Hydrocarbon compounds Chemical compounds containing only carbon and hydrogen.

Hydrocarbon-producing resource A resource such as coal and oil shale (kerogen) that produces derived hydrocarbons by the application of conversion processes; the hydrocarbons so produced are not naturally occurring materials.

Hydrocarbon reserves The volume of hydrocarbons measured at atmospheric conditions that will be produced economically by using any of the existing production methods at the date of evaluation.

Hydrocarbon resource A resource, such as petroleum or natural gas, that can produce naturally occurring hydrocarbons without the application of conversion processes.

Hydrocarbons Chemical compounds fully constituted by hydrogen and carbon.

Immiscibility The inability of two or more fluids to have complete mutual solubility; they co-exist as separate phases.

Immiscible carbon dioxide displacement The injection of carbon dioxide into an oil reservoir to effect oil displacement under conditions in which miscibility with reservoir oil is not obtained. See carbon dioxide augmented waterflooding.

Immiscible displacement A displacement of oil by a fluid (gas or water) that is conducted under conditions so that interfaces exist between the driving fluid and the oil.

Impurities and plant liquefiables shrinkage factor (iplsf) The fraction obtained by considering the non-hydrocarbon gas impurities (sulfur, carbon dioxide, nitrogen compounds, etc.) contained in the sour gas, in addition to shrinkage caused by the generation of liquids in gas processing plant.

Impurities shrinkage factor (isf) The fraction that results from considering the non-hydrocarbon gas impurities (sulfur, carbon dioxide, nitrogen compounds, etc.) contained in the sour gas. It is obtained from the operation statistics of the last annual period of the gas processing complex (GPC) that processes the production of the field analyzed.

Incompatibility The immiscibility of petroleum products and also of different crude oils, which is often reflected in the formation of a separate phase after mixing and/or storage.

Incremental ultimate recovery The difference between the quantity of oil that can be recovered by enhanced oil recovery methods and the quantity of oil that can be recovered by conventional recovery methods.

Infill drilling Drilling additional wells within an established pattern.

Initial boiling point The recorded temperature when the first drop of liquid falls from the end of the condenser.

Initial vapor pressure The vapor pressure of a liquid of a specified temperature and 100% evaporated.

Injection profile The vertical flow rate distribution of fluid flowing from the well bore into a reservoir.

Injection well A well in an oil field used for injecting fluids into a reservoir.

Injectivity The relative ease with which a fluid is injected into a porous rock.

In situ In its original place or in the reservoir.

In situ combustion An enhanced process consisting of injecting air or oxygen-enriched air into a reservoir under conditions that favor burning part of the in situ petroleum, advancing this burning zone, and recovering oil heated from a nearby producing well.

Instability The inability of a petroleum product to exist for periods of time without change to the product.

Integrity The maintenance of a slug or bank at its preferred composition without too much dispersion or mixing.

Interface The thin surface area separating two immiscible fluids that are in contact with each other.

Interfacial film A thin layer of material at the interface between two fluids that differs in composition from the bulk fluids.

Interfacial tension The strength of the film separating two immiscible fluids (e.g., oil and water or microemulsion and oil), measured in dynes (force) per centimeter or millidynes per centimeter.

Interfacial viscosity The viscosity of the interfacial film between two immiscible liquids.

Interference testing A type of pressure transient test in which pressure is measured over time in a closed-in well while nearby wells are produced; flow and communication between wells can sometimes be deduced from an interference test.

Interphase mass transfer The net transfer of chemical compounds between two or more phases.

Kaolinite A clay mineral formed by hydrothermal activity at the time of rock formation or by chemical weathering of rock with high feldspar content; it is usually associated with intrusive granite rock with high feldspar content.

Kerogen A complex carbonaceous (organic) material that occurs in sedimentary rock and shale; it is generally insoluble in common organic solvents. It produces hydrocarbons when subjected to a heat.

Kinematic viscosity The ratio of viscosity to density, both measured at the same temperature.

Kriging A technique used in reservoir description for interpolation of reservoir parameters between wells based on random field theory.

Light oil Oil with specific gravity of more than 27° API but less than or equal to 38° API.

Light petroleum Petroleum having an API gravity greater than 20°.

Limolite Fine-grain sedimentary rock that is transported by water. The granulometrics range from fine sand to clay.

Lithology The geological characteristics of the reservoir rock.

Lorenz coefficient A permeability heterogeneity factor.

Maltenes The fraction of petroleum that is soluble in, for example, pentane or heptane; it is also the term arbitrarily assigned to the pentane-soluble portion of petroleum that is relatively high boiling (>300°C, 760 mm). See also petrolenes. Also referred to as deasphaltened oil.

Marx-Langenheim model Mathematical equations for calculating heat transfer in a hot-water or steamflood.

MEOR microbial enhanced oil recovery.

Metagenesis The alteration of organic matter during the formation of petroleum that may involve temperatures above 200°C (390°F). See also catagenesis and diagenesis.

Metamorphic Rocks resulting from the transformation that commonly takes place at great depths due to pressure and temperature. The original rocks may be sedimentary, igneous, or metamorphic.

Mica A complex aluminum silicate mineral that is transparent, tough, flexible, and elastic.

Micellar fluid (surfactant slug) An aqueous mixture of surfactants, cosurfactants, salts, and hydrocarbons. The term micellar is derived from the word micelle, which is a submicroscopic aggregate of surfactant molecules and associated fluid.

Microemulsion A stable, finely dispersed mixture of oil, water, and chemicals (surfactants and alcohols).

Microemulsion or micellar/emulsion flooding An augmented waterflooding technique in which a surfactant system is injected in order to enhance oil displacement toward producing wells.

Microorganisms Animals or plants of microscopic size, such as bacteria.

Microscopic displacement efficiency The efficiency with which an oil-displacement process removes the oil from individual pores in the rock.

Middle-phase microemulsion A microemulsion phase containing a high concentration of both oil and water.

Migration (primary) The movement of hydrocarbons (oil and natural gas) from mature, organic-rich source rocks to a point where the oil and gas can collect as droplets or as a continuous phase of liquid hydrocarbon.

Migration (secondary) The movement of the hydrocarbons as a single, continuous fluid phase through water-saturated rocks, fractures, or faults followed by accumulation of the oil and gas in sediments (traps) from which further migration is prevented.

Mineral hydrocarbons Petroleum hydrocarbons, considered mineral because they come from the earth rather than from plants or animals.

Mineral oil The older term for petroleum. The term was introduced in the nineteenth century as a means of differentiating petroleum (rock oil) from whale oil, which was the predominant illuminant for oil lamps at the time.

Minerals Naturally occurring inorganic solids with well-defined crystalline structures.

Mineral seal oil A distillate fraction boiling between kerosene and gas oil.

Minimum miscibility pressure (MMP) see miscibility.

Miscibility An equilibrium condition, achieved after mixing two or more fluids, that is characterized by the absence of interfaces between the fluids. *First-contact miscibility* is miscibility in the usual sense, whereby two fluids can be mixed in all proportions without any interfaces forming. For example, at room temperature and pressure, ethyl alcohol and water are first-contact miscible. *Multiple-contact miscibility (dynamic miscibility)* is miscibility that is developed by repeated enrichment of one fluid phase with components from a second fluid phase with which it comes into contact. *Minimum miscibility* pressure is the minimum pressure above which two fluids become miscible at a given temperature, or can become miscible, by dynamic processes.

Miscible flooding See enhance oil recovery process.

Miscible fluid displacement (miscible displacement) An oil displacement process in which an alcohol, a refined hydrocarbon, a condensed petroleum gas, carbon dioxide, liquefied natural gas, or even exhaust gas is injected into an oil reservoir at pressure levels such that the injected gas or fluid and reservoir oil are miscible. The process may include the concurrent, alternating, or subsequent injection of water.

Mobility A measure of the ease with which a fluid moves through reservoir rock. It is the ratio of rock permeability to apparent fluid viscosity.

Mobility buffer The bank that protects a chemical slug from water invasion and dilution and assures mobility control.

Mobility control The ability to ensure that the mobility of the displacing fluid or bank is equal to or less than that of the displaced fluid or bank.

Mobility ratio The ratio of mobility of an injection fluid to mobility of fluid being displaced.

Modified alkaline flooding The addition of a cosurfactant and/or polymer to the alkaline flooding process.

Native asphalt See bitumen.

Natural gas A mixture of hydrocarbons existing in reservoirs in the gaseous phase or in solution in the oil, which remains in the gaseous phase under atmospheric conditions. It may contain some impurities or non-hydrocarbon substances (hydrogen sulfide, nitrogen, or carbon dioxide).

Natural gas liquids (NGL) The hydrocarbon liquids that condense during the processing of hydrocarbon gases that are produced from an oil or gas reservoir. See also natural gasoline.

Natural gasoline A mixture of liquid hydrocarbons extracted from natural gas suitable for blending with refinery gasoline.

Net thickness (hn) The thickness resulting from subtracting the portions of the reservoir that have no possibilities of producing hydrocarbon from the total thickness.

Non-associated gas The natural gas found in reservoirs that do not contain crude oil at the original pressure and temperature conditions.

Non-ionic surfactant A surfactant molecule containing no ionic charge.

Non-proved reserves Volumes of hydrocarbons and associated substances, evaluated at atmospheric conditions, resulting from the extrapolation of the characteristics and parameters of the reservoir beyond the limits of reasonable certainty or from the assumption of oil and gas forecasts with technical and economic scenarios other than those in operation or with a project in view.

Normal fault The result of the downward displacement of one of the strata from the horizontal. The angle is generally between 25 and 60 degrees and it is recognized by the absence of part of the stratigraphic column.

Observation wells Wells that are completed and equipped to measure reservoir conditions and/or sample reservoir fluids rather than to inject or produce reservoir fluids.

Oil The portion of petroleum that exists in the liquid phase in reservoirs and remains as such under original pressure and temperature conditions. Small amounts of non-hydrocarbon substances may be included. It has a viscosity of less than or equal to 10,000 centipoises at the original temperature of the reservoir, at atmospheric pressure, and gas-free (stabilized). Oil is commonly classified in terms of its specific gravity, and it is expressed in degrees API.

Oil breakthrough (time) The time at which the oil-water bank arrives at the producing well.

Oil equivalent (OE) The total of crude oil, condensate, plant liquids, and dry gas equivalent to liquid.

Oil originally in place (OOIP) The quantity of petroleum existing in a reservoir before oil-recovery operations begin.

Oil sand See tar sand.

Oil shale A fine-grained impervious sedimentary rock that contains an organic material called kerogen.

OOIP See oil originally in place.

Optimum salinity The salinity at which a middle-phase microemulsion containing equal concentrations of oil and water results from the mixture of a micellar fluid (surfactant slug) with oil.

Organic sedimentary rocks Rocks containing organic material such as residues of plant and animal remains/decay.

Original gas volume in place The amount of gas that is estimated to exist initially in the reservoir and that is confined by geologic and fluid boundaries, which may be expressed at reservoir or atmospheric conditions.

Original pressure The pressure prevailing in a reservoir that has never been produced. It is the pressure measured by a discovery well in a producing structure.

Original reserve The volume of hydrocarbons at atmospheric conditions that are expected to be recovered economically by using the exploitation methods and systems applicable at a specific date. It is a fraction of the discovered and economic reserve that may be obtained at the end of the reservoir exploitation.

Override The gravity-induced flow of a lighter fluid in a reservoir above another heavier fluid.

Pattern the horizontal pattern of injection and producing wells selected for a secondary or enhanced recovery project.

Pattern life The length of time a flood pattern participates in oil recovery.

Permeability The property of a rock to permit a fluid pass. It is a factor that indicates whether a reservoir has producing characteristics or not.

Petroleum A mixture of hydrocarbons composed of combinations of carbon and hydrogen atoms found in the porous spaces of rocks. Crude oil may contain other elements of a nonmetal origin, such as sulfur, oxygen, and nitrogen, in addition to trace metals as minor constituents. The compounds that form petroleum may be in a gaseous, liquid or solid state, depending on their nature and the existing pressure and temperature conditions.

Phase A separate fluid that coexists with other fluids. Gas, oil, water, and other stable fluids such as microemulsions are all called phases in enhanced oil recovery research.

Phase behavior The tendency of a fluid system to form phases as a result of changing temperature, pressure, or the bulk composition of the fluids or of individual fluid phases.

Phase diagram A graph of phase behavior. In chemical flooding, it is a graph that shows the relative volume of oil, brine, and sometimes one or more microemulsion phases. In carbon dioxide flooding, it shows conditions for formation of various liquid, vapor, and solid phases.

Phase properties Types of fluids, compositions, densities, viscosities, and relative amounts of oil, microemulsion, or solvent, and water formed when a micellar fluid (surfactant slug) or miscible solvent (e.g., CO_2) is mixed with oil.

Phase separation The formation of a separate phase that is usually the prelude to coke formation during a thermal process. This separate phase forms as a result of the instability/incompatibility of petroleum and petroleum products.

Physical limit The limit of the reservoir defined by any geological structures (faults, unconformities, change of facies, crests, and bases of formations, etc.), caused by contact between fluids or by the reduction to critical porosity of permeability limits or by the compound effect of these parameters.

Pilot project A project that is being executed in a small representative sector of a reservoir where tests performed are similar to those that will be implemented throughout the reservoir. The purpose is to gather information and/or obtain results that could be used to generalize an exploitation strategy in the oil field.

PINA analysis A method of analysis for paraffins (P), *iso*-paraffins (I), naphthenes (N), and aromatics (A).

PIONA analysis A method of analysis for paraffins (P), *iso*-paraffins (I), olefins (O), naphthenes (N), and aromatics (A).

Pitch The nonvolatile, brown-to-black, semi-solid-to-solid viscous product from the destructive distillation of many bituminous or other organic materials, especially coal.

Plant liquefiables shrinkage factor (plsf) The fraction arising from considering the liquefiables obtained in transportation to the processing complexes.

Plant liquids Natural gas liquids recovered in gas processing complexes, mainly consisting of ethane, propane, and butane.

Plant liquids recovery factor (plrf) The factor used to obtain the liquid portions recovered in the natural gas processing complex. It is obtained from the operation statistics of the last annual period of the gas processing complex that processes the production of the field analyzed.

Play A group of fields sharing geological similarities where the reservoir and the trap control the distribution of oil and gas.

Polymer In enhanced oil recovery, any very-high-molecular-weight material that is added to water to increase viscosity for polymer flooding.

Polymer augmented waterflooding Waterflooding in which organic polymers are injected with the water to improve horizontal and vertical sweep efficiency.

PONA analysis A method of analysis for paraffins (P), olefins (O), naphthenes (N), and aromatics (A).

Pore volume The total volume of all pores and fractures in a reservoir or part of a reservoir. It is also applied to catalyst samples.

Porosity The ratio between the pore volume existing in a rock and the total rock volume. It is a measure of a rock's storage capacity, the percentage of rock volume available to contain water or other fluid.

Possible reserves The volume of hydrocarbons that are less likely to be commercially recoverable than probable reserves, as suggested by the analysis of geological and engineering data.

Potential reserves Reserves based upon geological information about the types of sediments where such resources are likely to occur. They are considered to represent an educated guess.

Pour point The lowest temperature at which oil will pour or flow when it is chilled without disturbance under definite conditions.

Preflush A conditioning slug injected into a reservoir as the first step of an enhanced oil recovery process.

Pressure cores Cores cut into a special coring barrel that maintains reservoir pressure when brought to the surface. This prevents the loss of reservoir fluids that usually accompanies a drop in pressure from reservoir to atmospheric conditions.

Pressure gradient The rate of change of pressure with distance.

Pressure maintenance Augmenting the pressure (and energy) in a reservoir by injecting gas and/or water through one or more wells.

Pressure pulse test A technique for determining reservoir characteristics by injecting a sharp pulse of pressure in one well and detecting it in surrounding wells.

Pressure transient testing Measuring the effect of changes in pressure at one well on other wells in a field.

Primary recovery The extraction of petroleum by only using the natural energy available in the reservoirs to displace fluids through the reservoir rock to the wells.

Primary tracer A chemical that, when injected into a test well, reacts with reservoir fluids form a detectable chemical compound.

Probable reserves Nonproved reserves that are more likely to be commercially recoverable than not, as suggested by the analysis of geological and engineering data.

Producibility The rate at which oil or gas can produced from a reservoir through a well bore.

Producing well A well in an oil field used for removing fluids from a reservoir.

Prospective resource The amount of hydrocarbons evaluated at a given date of accumulations not yet discovered but which have been inferred and are estimated as recoverable.

Protopetroleum A generic term used to indicate the initial product formed from the precursors of petroleum.

Proved area The plant projection of the known part of the reservoir corresponding to the proved volume.

Proved reserves The volume of hydrocarbons or associated substances evaluated at atmospheric conditions that, by analysis of geological and engineering data, may be estimated with reasonable certainty to be commercially recoverable from a given date forward, from known reservoirs and under current economic conditions, operating methods, and government regulations. Such volume consists of the developed proved reserve and the undeveloped proved reserve.

Pulse-echo ultrasonic borehole televiewer A well-logging system wherein a pulsed, narrow acoustic beam scans the well as the tool is pulled up the bore hole; the amplitude of the reflecting beam is displayed on a cathode-ray tube, resulting in a pictorial representation of wellbore.

Quadrillion 1×10^{15}

Raw materials Minerals extracted from the earth prior to any refining or treating.

Recovery factor (rf) The ratio between the original volume of oil or gas at atmospheric conditions and the original reserves of the reservoir.

Refinery A series of integrated unit processes by which petroleum can be converted to a slate of useful (salable) products.

Regression A geological term used to define the elevation of one part of the continent over sea level, as a result of the ascent of the continent or the lowering of the sea level.

Relative permeability The capacity of a fluid, such as water, gas, or oil, to flow through a rock when it is saturated with two or more fluids. The value of the permeability of a saturated rock with two or more fluids is different to the permeability value of the same rock saturated with just one fluid.

Remaining reserves The volume of hydrocarbons measured at atmospheric conditions that are still to be commercially recoverable from a reservoir at a given date, using the applicable exploitation techniques. It is the difference between the original reserve and the cumulative hydrocarbon production at a given date.

Reserve replacement rate A rate that indicates the amount of hydrocarbons replaced or incorporated by new discoveries compared with what has been produced in a given period. It is the coefficient that arises from dividing the new discoveries by production during the period of analysis. It is generally referred to in annual terms and is expressed as a percentage.

Reserve-production ratio The result of dividing the remaining reserve at a given date by the production in a period. This indicator assumes constant production, hydrocarbon prices, and extraction costs, without variation over time, in addition to the nonexistence of new discoveries in the future.

Reserves Well-identified resources that can be profitably extracted and utilized with existing technology.

Reservoir The portion of the geological trap containing hydrocarbons that acts as a hydraulically interconnected system, where the hydrocarbons are found at an elevated temperature and pressure occupying the porous spaces.

Reservoir simulation The analysis and prediction of reservoir performance with a computer model.

Residual resistance factor The reduction in permeability of rock to water caused by the adsorption of polymer.

Resistance factor A measure of resistance to flow of a polymer solution relative to the resistance to flow of water.

Resource The total volume of hydrocarbons existing in subsurface rocks. It is also known as original in-situ volume.

Retention The loss of chemical components due to adsorption onto the rock's surface, to precipitation, or to trapping within the reservoir.

Reverse fault The result of compression forces where one of the strata is displaced upwards from the horizontal.

Revision The reserve resulting from comparing the previous year's evaluation with the new one, in which new geological, geophysical, operation, and reservoir performance information is considered, in addition to variations in hydrocarbon prices and extraction costs. It does not include well drilling.

Rock matrix The granular structure of a rock or porous medium.

Run-of-the-river reservoirs Reservoirs with a large rate of flow-through compared to their volume.

Salinity The concentration of salt in water.

Sand A course granular mineral mainly composed of quartz grains that is derived from the chemical and physical weathering of rocks rich in quartz, notably sandstone and granite.

Sand face The cylindrical wall of the well bore, through which the fluids must flow to or from the reservoir.

Sandstone A sedimentary rock formed by compaction and cementation of sand grains. It can be classified according to the mineral composition of the sand and cement.

SARA analysis A method of fractionation by which petroleum is separated into saturates (S), aromatics (A), resins (R), and asphaltene (A) fractions.

Saturation The ratio of the volume of a single fluid in the pores to pore volume, expressed as a percent and applied to water, oil, or gas separately. The sum of the saturations of each fluid in a pore volume is 100 percent.

Saturation pressure The pressure at which the first gas bubble is formed, when it goes from the liquid phase to the two-phase region.

Saybolt Furol viscosity The time, in seconds (Saybolt Furol Seconds, SFS), for 60 ml of fluid to flow through a capillary tube in a Saybolt Furol viscometer at specified temperatures between 70°F and 210°F. The method is appropriate for high-viscosity oils such as transmission, gear, and heavy fuel oils.

Saybolt Universal viscosity The time, in seconds (Saybolt Universal Seconds, SUS), for 60 ml of fluid to flow through a capillary tube in a Saybolt Universal viscometer at a given temperature.

Screening guide A list of reservoir rock and fluid properties critical to an enhanced oil recovery process.

Secondary recovery Techniques used for the additional extraction of petroleum after primary recovery. This includes gas or water injection, partly to maintain reservoir pressure.

Secondary tracer The product of the chemical reaction between reservoir fluids and an injected primary tracer.

Sediment An insoluble solid formed as a result of the storage instability and/or the thermal instability of petroleum and petroleum products.

Sedimentary Formed by or from deposits of sediments, especially from sand grains or silts transported from their source and deposited in water (as sandstone and shale) or from calcareous remains of organisms (as limestone).

Sedimentary strata These strata typically consist of mixtures of clay, silt, sand, organic matter, and various minerals; they are formed by or from deposits of sediments, especially from sand grains or silts transported from their source and deposited in water (as sandstone and shale) or from calcareous remains of organisms (as limestone).

Seismic section A seismic profile that uses the reflection of seismic waves to determine the geological subsurface.

Slime A name used for petroleum in ancient texts.

Sludge A semi-solid-to-solid product which results from the storage instability and/or the thermal instability of petroleum and petroleum products.

Slug A quantity of fluid injected into a reservoir during enhanced oil recovery.

Sonic log A well log based on the time required for sound to travel through rock. It is useful in determining porosity.

Sour crude oil Crude oil containing an abnormally large amount of sulfur compounds. See also sweet crude oil.

Spacing The optimum distance between hydrocarbon producing wells in a field or reservoir.

Specific gravity An intensive property of matter that is related to the mass of a substance and its volume through the coefficient between these two quantities. It is expressed in grams per cubic centimeter or in pounds per gallon.

Standard conditions The reference amounts for pressure and temperature. In the English system, they are 14.73 pounds per square inch for the pressure and 60°F for temperature.

Steam distillation Distillation in which vaporization of the volatile constituents is effected at a lower temperature by introduction of steam (open steam) directly into the charge.

Steam drive injection (steam injection) An enhanced oil recovery process in which steam is continuously injected into one set of wells (injection wells) or other injection source to effect oil displacement toward and production from a second set of wells (production wells). Steam stimulation of production wells is *direct steam stimulation,* whereas steam drive by steam injection to increase production from other wells is *indirect steam stimulation.*

Steam stimulation The injection of steam into a well and the subsequent production of oil from the same well.

Stiles method A simple approximate method for calculating oil recovery by waterflood that assumes separate layers (stratified reservoirs) for the permeability distribution.

Stimulation The process of acidifying or fracturing carried out to expand existing ducts or to create new ones in the source rock formation.

Strata Layers including the solid iron-rich inner core, molten outer core, mantle, and crust of the earth.

Stratigraphy The part of geology that studies the origin, composition, distribution, and succession of rock strata.

Stripper well A well that produces (strips from the reservoir) oil or gas.

Structural nose A term used in structural geology to define a geometric form protruding from a main body.

Sucker rod pumping system A method of artificial lift in which a subsurface pump located at or near the bottom of the well and connected to a string of sucker rods is used to lift the well fluid to the surface.

Super-light oil Oil with a specific gravity more than 38° API.

Surface active material A chemical compound, molecule, or aggregate of molecules with physical properties that cause it to adsorb at the interface between *two* immiscible liquids, resulting in a reduction of interfacial tension or the formation of a microemulsion.

Surfactant A type of chemical that reduces interfacial resistance to mixing between oil and water or changes the degree to which water wets reservoir rock.

Sweet crude oil Crude oil containing little sulfur. See also sour crude oil.

Sweetening plant An industrial plant used to treat gaseous mixtures and light petroleum fractions in order to eliminate undesirable or corrosive sulfur compounds to improve their color, odor, and stability.

Swelling An increase in the volume of crude oil caused by absorption of enhanced oil recovery fluids, especially carbon dioxide. It is also an increase in volume of clays when exposed to brine.

Swept zone The volume of rock that is effectively swept by injected fluids.

Synthetic crude oil (syncrude) A hydrocarbon product produced by the conversion of coal, oil shale, or tar sand bitumen that resembles conventional crude oil. It can be refined in a petroleum refinery.

Tar The volatile, brown-to-black, oily, viscous product from the destructive distillation of many bituminous or other organic materials, especially coal. It is a name used for petroleum in ancient texts.

Tar sand See bituminous sand.

Technical reserves The accumulative production derived from a production forecast in which economic criteria are not applied.

Thermal recovery See enhanced oil recovery process.

Thief zone Any geologic stratum not intended to receive injected fluids in which significant amounts of injected fluids are lost. Fluids may reach the thief zone due to an improper completion or a faulty cement job.

Time-lapse logging The repeated use of calibrated well logs to quantitatively observe changes in measurable reservoir properties over time.

Total thickness (h) The thickness from the top of the formation of interest down to a vertical boundary determined by a water level or by a change of formation.

Tracer test A technique for determining fluid-flow paths in a reservoir by adding small quantities of easily detected material (often radioactive) to the flowing fluid and monitoring their appearance at production wells. It is also used in cyclic injection to appraise oil saturation.

Transgression The geological term used to define the immersion of one part of the continent under sea level as a result of a descent of the continent or an elevation of the sea level.

Transmissibility (transmissivity) An index of producibility of a reservoir or zone; it is the product of permeability and layer thickness.

Transport liquefiables shrinkage factor (tlsf) The fraction obtained by considering the liquefiables obtained in transportation to the processing complexes.

Trap Geometry that permits the concentration of hydrocarbons. It also refers to sediment in which oil and gas accumulate, from which further migration is prevented.

Triaxial borehole seismic survey A technique for detecting the orientation of hydraulically induced fractures, wherein a tool holding three mutually seismic detectors is clamped in the borehole during fracturing; fracture orientation is deduced through analysis of the detected microseismic perpendicular events that are generated by the fracturing process.

Trillion 1×10^{12}

Ultimate analysis Elemental composition.

Ultimate recovery The cumulative quantity of oil that will be recovered when revenues from further production no longer justify the costs of the additional production.

Unconformity A surface of erosion that separates younger strata from older rocks.

Undeveloped proved area The plant projection of the extension drained by the future producing wells of a producing reservoir and located within the undeveloped proved reserve.

Undeveloped proved reserves The volume of hydrocarbons that is expected to be recovered through wells without current facilities for production or transportation and future wells. This category may include the estimated reserve of enhanced recovery projects, with pilot testing or with the recovery mechanism proposed in operation that has been predicted with a high degree of certainty in reservoirs that benefit from this kind of exploitation.

Undiscovered resource The volume of hydrocarbons with uncertainty but whose existence is inferred in geological basins through favorable factors resulting from the geological, geophysical and geochemical interpretation. They are known as prospective resources when considered commercially recoverable.

Universal viscosity See Saybolt Universal viscosity.

Upper-phase microemulsion A microemulsion phase containing a high concentration of oil that, when viewed in a test tube, resides on top of a water phase.

Vertical seismic profiling (VSP) A method of conducting seismic surveys in the bore hole for detailed subsurface information.

Vertical sweep efficiency The fraction of the layers or vertically distributed zones of a reservoir that are effectively contacted by displacing fluids.

Visbreaking A process for reducing the viscosity of heavy feedstocks by controlled thermal decomposition.

Viscosity A measure of the ability of a liquid to flow or a measure of its resistance to flow. It is the force required to move a plane surface of area 1 m^2 over another parallel plane surface 1 m away at a rate of 1 m/sec when both surfaces are immersed in the fluid.

Volumetric sweep The fraction of the total reservoir volume within a flood pattern that is effectively contacted by injected fluids.

Vertical seismic profiling (VSP) A method of conducting seismic surveys in the bore hole for detailed subsurface information.

Waterflood The injection of water to displace oil from a reservoir (usually a secondary recovery process).

Waterflood mobility ratio The mobility ratio of water displacing oil during waterflooding. See also mobility ratio.

Waterflood residual The waterflood residual oil saturation; it is the saturation of oil remaining after waterflooding in those regions of the reservoir that have been thoroughly contacted by water.

Well abandonment The final activity in the operation of a well when it is permanently closed under safety and environment preservation conditions.

Well bore The hole in the earth comprising a well.

Well completion The complete outfitting of an oil well for either oil production or fluid injection. It is also the technique used to control fluid communication with the reservoir.

Wellhead The portion of an oil well above the surface of the ground.

Well logs The information concerning subsurface formations obtained by means of electric, acoustic, and radioactive tools inserted in the wells. The log also includes information about drilling and the analysis of mud and cuts, cores, and formation tests.

Wet gas A mixture of hydrocarbons obtained from processing natural gas from which non-hydrocarbon impurities or compounds have been eliminated and whose content of components that are heavier than methane is such that it can be commercially processed.

Wettability The relative degree to which a fluid will spread on (or coat) a solid surface in the presence of other immiscible fluids.

Wettability number A measure of the degree to which a reservoir rock is water wet or oil wet, based on capillary pressure curves.

Wettability reversal The reversal of the preferred fluid wettability of a rock, e.g., from water wet to oil wet or vice versa.

INDEX

F

G

P

T